STARTING A TECHNOLOGY BUSINESS

STARTING A TECHNOLOGY BUSINESS

John Allen
Ph.D., B.Sc., FRSC, C.Chem., FIFST

Published in association with
NatWest Technology Unit

Pitman Publishing
128 Long Acre, London WC2E 9AN

A Division of Longman Group UK Limited

First published in 1992

British Library Cataloguing in Publication Data
A CIP catalogue record for this book can be obtained from the British Library.

ISBN 0 273 03671 8

Typeset, printed and bound in Great Britain

CONTENTS

PREFACE

When I was first asked to set up the Technology Unit, which has since established NatWest as the leading bank for technology businesses, I quickly appreciated that one of the biggest problems was obtaining advice on how to run a business in this unique sector. I could find books on starting a small business generally, but none on the issues I should face, were I to set up a technology business.

There is no doubt that technology-based businesses have to take into consideration all the elements vital to the running of a successful business, such as developing a sound strategy, planning and financial control, marketing and selling, and a strong management team to pull all the elements together. In addition, however, technology entrepreneurs need to focus their minds on the protection of their intellectual property, whether by patenting, copyrights, design rights or trademarks, and the best way of taking their product to market – sole manufacture, licensing or joint venture.

The needs of such businesses are therefore more complex. In addition, they are competing in a much faster moving marketplace; also, as a result of spending more time on research and development, longer lead-in times are required before income can be generated and profitability achieved.

NatWest recognized the importance of the technology sector some years ago. Having set up the Technology Unit in May 1989, I immediately put in place a UK-wide network of trained Technology Managers to understand and support the needs of technology-based businesses on a local basis, together with embarking on a number of other initiatives to provide a complete infrastructure of support.

My research in the early days, and ever since, has shown that what these businesses need, often more than finance, is ongoing support and guidance. Banks are now beginning to meet this need – and not before time. Certainly the UK has not had a lack of truly innovative ideas, but to date the support to help the commercialisation process has been patchy, to say the least.

It is with all these factors in mind that I decided to commission this book, which addresses all the issues facing a technology business in the 1990s and beyond. Hopefully, its long-awaited arrival should enable entrepreneurs to avoid many of the pitfalls, to make the right choices towards commercializing their ideas, and to get onto the road to commercial success.

Duncan Matthews
Senior Manager
NatWest Technology Unit

FOREWORD

I wish to thank Duncan Matthews and Peter Ives of the NatWest Technology Unit for the invitation to write this book. The idea was thus theirs originally, but I must take full responsibility for any shortcomings and the views expressed.

NatWest, having established a group to support technology-based business, were concerned that there was precious little in the way of guidance for the smaller firm. Although station and airport bookstalls have shelves packed with books on how to run IBM and – at the other extreme – how to start a print shop or a taxi service, there was nothing that focused on the particular opportunities and problems of the technology-based business.

It is my hope that this book addresses these points in a realistic manner, and that it will be of use to the entrepreneur in the development of his or her business. I also hope it will stimulate the potential investor to take more interest in technology-based proposals, or at least to initiate a dialogue between themselves and the entrepreneurs and their advisers.

To the outsider, technology-based businesses can seem exciting and somewhat glamorous. The idea of developing and making novel products that depend on the application of new research can appear an attractive way to make a living. The media enjoy publicising stories of brilliant innovators with ingenious ideas, and give the impression that a rosy future inevitably lies in front of them.

To the investor, technology-based projects can seem risky. The inventors frequently lack managerial experience, the products are untried, the market appears uncertain and the amount of cash needed to develop the concept to a prototype and then into production can escalate rapidly, way beyond any initial predictions.

To those of us assisting such firms through their formative stages, the major difficulties are not so different from those faced by any business. The biggest problem is management, and the need to bring appropriate managerial skills, by whatever means, to technology-based businesses is paramount. For this reason, the topic of management is a sub-theme to virtually the whole of the book.

Yet the very future of our national economy depends on the maintenance of our competitiveness. Our ability to invent *and then exploit the invention efficiently* are critical factors, and it is clear from the experience of many countries that smaller firms can be an effective vehicle for this. The well-managed smaller firm can be more flexible, have a sharper focus and be less constrained by rigid administrative procedures than large companies.

Such firms, however, find it a challenge to grow in the best of times, and in an adverse economic climate even the better ones can run into difficulties. It is vital to develop a more practical and realistic supporting network to help such firms, involving Government, the financial establishment, management advisers and sources of technology such as the universities. The barriers to exploitation, such as the Government's refusal to fund 'near-market' research and the misplaced aversion to technology by many bankers and venture capitalists, should be reduced by discussion and reassessment.

If I may be allowed to ride my own particular hobby horse, support for Science Parks, which have a successful track record of supporting innovation over the past decade, should be increased. By providing an attractive location, with management support available and with a close liaison with a university or other centre of technology, they provide an excellent environment for the technology-based company to flourish.

I am in no way advocating support for 'lame ducks'. I believe in incubators, not life-support machines! Even the best-managed companies face an uphill struggle to become viable in the current difficult world economy and against strong and increasing competition from other countries.

I must now record my thanks to several others for their help. I am particularly grateful to Mrs. Jacqueline Needle of W.H. Beck Greener & Co. for her expert help over the complexities of intellectual property rights; she was kind enough to check the facts, but I still take responsibility for the opinions. Douglas McQueen of ETH in Zürich has allowed me to crib his scheme of the flow of technology between industry and academia. Alan Toop of The Sales Machine gave permission for me to quote from an article about cash flow forecasts which I found ideal; and Dr. John Slack of Aston Molecules kindly allowed me to give his company's history as an example.

Finally, I must acknowledge my colleagues in the UK Science Parks Association (UKSPA) and the European Business Innovation Centre Network (EBN). Their experience in supporting technology-based businesses is second to none, and I am glad to be associated with them.

John C. Allen.

Chapter 1
TECHNOLOGY-BASED BUSINESSES

What is a technology-based business

This book is principally concerned with helping you, the entrepreneur, to turn your technology-based business idea into a commercial success. It is not a theoretical or comprehensive study of innovative business, but a practical guide to the major factors which can determine success.

Since businesses ultimately depend on people rather than things or systems, it is important to allow for this fact. Some books give the impression that, to ensure fame and fortune, you only have to follow the 'recipe' given. This is totally false. I will try to describe the important ingredients, to explain why they are important and to advise how best to put them together. However, it will be up to you, the entrepreneur, to make the business successful. Almost invariably, your personal qualities will play the major role in this.

To start, however, some general definitions and descriptions are essential. The first word I want to focus on is 'success'.

■ THE CRITERION OF SUCCESS

By 'success', I mean 'profitable'. It is always interesting to see a new application of a scientific principle in a commercial product, and there are many prestigious prizes awarded for this. Are the winners of such competitions therefore inevitably successful in the business world?

Some do indeed have long-term commercial successes, but the organizers of well-publicised national competitions have been somewhat embarrassed to discover that technological ingenuity and even apparent industrial applicability by no means guarantees financial success. On following up the progress of prize-winners from previous years, it was revealed that many of the inventors had progressed no further, had not found a financial backer, were making painfully slow progress or had simply gone bust.

There are a number of reasons for this and they will be dealt with in detail in subsequent chapters, but the principal cause behind the inordinate number of crashes was the failure to appreciate that innovation, indeed, *any* business venture, must be market-led.

In the present context, suffice it to say that 'success' means making the cash registers ring, so that a profit is produced. The technology is but a means to that end.

■ THE TECHNOLOGY CONTENT

Next, it is worth considering what I mean by the term 'technology-based business'. There are quite a few postgraduate student projects which have split semantic hairs over the definitions of technology businesses, producing in the process a litter of divisions and sub-divisions of the term 'technology'. Chapter 2 bows a little in this direction by discussing the *flow* of technology in some detail, but I intend to take a purely operational and pragmatic approach to the definition as applied to the sort of businesses envisaged and run by the entrepreneurs this book addresses.

By a 'technology-based' business I mean *a business whose products or services depend to a significant extent on the application of scientific or technological skills or knowledge*. (Whether it be a novel application of advanced technology to provide a totally new product or service, or an application of existing technology in an innovative manner.)

I would exclude large assembly plants, such as factories assembling video recorders which are designed elsewhere. The product is certainly technology-based, but all the technology is imported. On the other hand, if an electronics technician started a small company to make some novel device which, say, made it much easier to program the recorder, as a manufacturing business it *would* fall within my definition.

The products of technology-based businesses need not be technological. There is a factory in North Wales which makes high-quality, ready-cooked meals principally for the catering trade, whose existence largely depends on the use of a totally novel form of ohmic heating. This highly-efficient and cost-effective heating *process*, developed over the past few years by a national research establishment, is an integral and essential part of the business.

Perhaps these examples help to delineate the boundaries of this operational definition. The bulk of technology-based businesses, however, are what most of us imagine them to be. They are firms developing and selling in areas such as software, computer applications and peripherals, robots, electronic and telecommunications products, biotechnology, pharmaceutical and health-care products, new materials, alternative energy production or semiconductors. (All these and more are dealt with in Chapter 2.)

They are also companies providing consultancy and services in all these areas, sometimes revolutionizing the production of a traditional or mundane product.

□ A local company making multi-pack wrapping for supermarket products licensed in a new piece of technology from Germany. The invention, still in a fairly embryonic state, was a wrapping which was cheaper, easier for the customer to pick up and carry and was 'greener' on two counts. Firstly, it used only biodegradable materials, and secondly it used far less energy than does the formation of the typical 'shrink-wrap' pack.

It took considerable foresight on behalf of the UK company to see the long-term advantages and the market to be gained, and it took nearly two years, substantial technical and design input and a considerable amount of cash to develop the invention into a reliable working production line for the new wrapping. At the present, the company has had a number of major successes and the product is currently securing a significant and growing share of a large market. □

The essence of the technology based business is therefore the translation of research and technology into marketable products or services. The novelty and quality of the science involved is very important, but the crucial criteria are more frequently the existence of a market and the quality of the management. This is why, out of the ten chapters in this book, only one is devoted to technology and the rest to the business aspects. After all, *you* are the technologist with the know-how and the bright idea.

The originators of technology-based businesses

■ THE NON-TECHNICAL

Entrepreneurs wishing to start technology-based businesses come in all shapes and sizes. Most of them know something about technology, but even that is not invariably so.

Sometimes a person who regularly *uses* a piece of technology but who knows little about the technical detail (a musician using a synthesiser or a word-processor operator, for instance) will see the market for a new peripheral or extension of the machine's capability which the technical expert will completely miss.

Since the market for the product is recognized from the outset, this can be an extremely important source of new ideas when it occurs. The next stage is to inject the necessary technology to develop the product into a working prototype.

■ THE ACADEMIC

Most innovators have got some kind of scientific or technical training, experience or know-how; and most innovative ideas are derived or 'spun off' from work done in larger institutions, be they academic or industrial. The academic world is an important spawning ground for new exploitable ideas, and an increasing number of university and polytechnic staff are deciding (and are being allowed the opportunity) to extend the research and consultancy work they undertake to the point of forming a company.

Several hundred such companies are in existence on UK Science Parks alone. Most parks are partly or wholly managed by the university or academic institution, and many welcome such 'spin offs'. Some also have arrangements to allow the member of staff some flexibility and freedom to develop his commercial ideas, and view such activity as bringing ultimate benefit to the university.

Since the move towards commercialization can commence from within the academic environment, there is a good opportunity to try out some of the facets of business life before a definite commitment is made. It is possible to develop contacts with potential customers, gauge the size of the market and actually engage in some consultancy work.

This is a relatively 'fail-safe' position to be in, but it can be deluding. It is easy to think that this is all there is to running a business. If you decide to 'go it alone', you must consider:

Can you continue to innovate?
At present, your projected 'spin off' probably depends on just one product. Even the most novel product has a limited lifetime: are there others to take its place? And are the others coming from you or your competitors?

Can you afford the costs of developing these?
The first product may well have come from your work in the academic world, and its development was probably supported, at least in part, by use of facilities and equipment owned by the university or research institute. When you have your own company, will you have to pay for all these, and will you be able to afford it?

Who has the rights to the intellectual property in your invention?
Chapter 3 considers this in more detail, but the legal ownership of any invention needs to be clear before embarking on a business which depends on it. Perhaps the university or research institution has a policy on this? You need to make very sure.

Working on a little consultancy or commercial activity from within the sheltered confines of a university is far removed from 'being out there on your own', with a new business, a substantial overdraft and only your own skills and hard work between success and disaster. On the other hand, many find it invigorating, enjoy the challenge and relish the thought of their independence and the subsequent financial rewards.

Technicians in academic departments or research institutions, frequently conceive original business ideas. These are often in the service sector, providing highly specialized support services such as maintenance of electron microscopes, computers and satellite communications equipment. The academic or research world from which they come often provides the market for their services, and they usually know it well beforehand. They may even know most of the individual customers.

You may have a good commercial idea, have been uncertain of what to do and now having read this far, be convinced that you should shelve any further thoughts of moving out from the academic world to start a business. What then should you do?

One route is to first establish your right to the intellectual property and then enter into a licensing agreement with another company to make, market and sell the product under licence (*see* Chapter 3). This is neither easy nor rapid, and it is important to enlist the services of a licensing agent and a solicitor who is familiar with this sort of activity: not all are. Ensure that, when entering into discussions with potential licensees, a Confidentiality Agreement is first signed by both parties. There is a specimen draft agreement at the end of Chapter 3.

■ THE INDUSTRIALIST

The entrepreneur coming from an industrial environment has the advantage of knowing something about industry, the costs involved, the constraints and the need to maintain strict commercial deadlines. Even the industrial research scientist will be aware of the various pressures to meet targets and beat the competition.

Nevertheless, the industrial entrepreneur faces greater disadvantages than does the academic. The typical company may not take too kindly to allowing the time involved in setting up a new business. Since the product may be an extension of one of their own, or, in the extreme case, be in direct competition, they may place restrictions on the whole process. It is vital to carefully check the conditions of contract of employment and other legal aspects.

Another factor is that, despite a more general familiarity with industry, the entrepreneur will not possess the overall management skills required to start and run a business. A technical manager may be first-rate at the technology but will not perhaps know how to prepare or execute a marketing plan. A cash-flow analysis may be a closed book to even an experienced production manager. Planning a production schedule, which demands constant attention to optimize the efficiency of operation, may be a nightmare to a skilled researcher who is more interested in overcoming technical difficulties.

It is not uncommon for new technology-based businesses to be projects devised by a small *team* of entrepreneurs, usually two or three people. These often come from the same academic department, the same company or the same laboratory or workshop.

It is therefore important to ask a very personal question at this juncture, to the individual and to the entrepreneur team: *have you got what it takes*?

■ HAVE YOU GOT WHAT IT TAKES?

In fact you probably have, or you would not be motivated enough to read this far. The important thing to realize is, when starting your own business, if *you* haven't got what it takes then no one else has, because you only have your own experience, skills and industry to rely on for most of the time.

What are your strengths? It may be that you have a unique technical skill and a marketable idea to match. You may, through your present occupation, have developed both knowledge of the market and contacts with potential customers. You may have management skills through running a research team (no mean task!) or a production group.

On the other hand, you may not want to spend increasing amounts of time on management of a company, however committed you are to seeing your concepts turn into marketable products. It may well be the wise decision to concentrate on the technical side of the business and produce the next product.

Also, do you actually know what you want to sell? This may seem a silly thing to say, but there is a type of inventor who has a good idea but is totally mentally incapable of fixing on a particular product with a specific price tag to serve a defined market.

I was once faced by such a man with a novel idea of exploiting a concept to enable low-cost image processing. 'What product do you expect to make, who will you sell it to and how many do you expect to sell?' was my next question.

'Oh, the actual product doesn't matter' was the reply. 'My idea has such a general application that there are dozens of things it can be used for.' Only after some time and much effort was he convinced that the R&D was not the only thing which mattered, and that the world would not gladly beat a path to his doorstep once the validity of the concept had been proven.

A subtle variation on this which is sometimes encountered is what Harry Fitzgibbons of the Hambros Advanced Technology Trust dubbed the 'endless development syndrome'. This occurs when an innovator with an excellent product which could well satisfy the customer can never quite bring himself to stop improving and developing it, even temporarily, to put it into production and sell a few.

☐ Harry illustrates this well with the story of 'Joe's Tilting Table'. Joe was a committed innovator living in the North of England who had designed and patented a multi-axis tilting table. He had built a prototype of a six-axis machine, and had had assistance from the local university in writing the necessary controlling software. It was clearly a good invention, and he had been a finalist in the Prince of Wales Innovation Award. He had even sold the house he built himself to fund the development, and then received support from Hambros, principally to build three machines to fulfil three 'test' orders from major and prestigious companies.

However, in Joe's view the table was not ready. He felt he must improve the accuracy and enhance the functionality of his invention before it could possibly be put on the market, although, when pressed, he acknowledged that the customers did not really want these improvements and that the company desperately needed to generate cash flow.

Hambros came to the eventual conclusion that Joe would be 'better off entirely in charge of his destiny', and that their involvement in the project should cease. The project seemed destined to be in a perpetual state of further development. ☐

Of equal importance to your strengths are your weaknesses. You may have a marketable idea without the specialized technical skill to make a prototype and develop a product. You may have the prototype but lack the experience to manufacture the product in the quantity or quality required. You may lack the necessary financial skills to successfully prepare a Business Plan (*see* Chapter 4 p. 86) and control the finances of a cash-hungry R&D company.

It may well be, however, that your strengths are quite sufficient to launch the business you have in mind. For instance, a typical small, specialized consultancy will not require the range of management skills of a production company. One suitable person may be able to manage the whole operation at its initial level.

However, for a production company – what is sometimes termed a 'hard' company – a broader range and depth of management skills are necessary. The management requirements for different categories of technology-based companies are considered on p. 21–25.

In some instances learning can be done by experience 'on the job', supplemented by suitable training. For example, the entrepreneur consultant who wishes to learn something of marketing or financial skills can acquire them by undertaking a suitable course of lectures or, nowadays, distance learning. If you feel you are in this position, talk to your local Training and Enterprise Council, Enterprise Agency, Innovation Centre or Science Park. They will advise you on what is available and most suitable for your needs.

However, remedying a management deficiency by training frequently takes too long or is ineffective: for instance, not every technical expert *wants* to learn about marketing, and such a person is best left to contribute his own special skills to the company. In any case, it can take several years to acquire sufficient skills to become a competent general manager of even a small production company. It therefore becomes imperative for the present management to supplement themselves with someone who can fill the gap. This in turn means that the management must first find somebody and then be prepared to give up some of the company to him.

Any supplement to the management team will want a share of the company. To a pair of researchers who have used their highly specialized knowledge researching and developing a totally new product, or a couple of industrial technicians who have spent long hours planning their new business, this can be a difficult decision.

However, the choice is clear. The team can go ahead as it is, with the knowledge that it is seriously deficient in a key management skill, with all the problems of raising capital and developing the business which that entails; the chances are that the business will fail and the entrepreneurs will lose their investment and opportunity. Or they can lose some equity, bring in the skill needed and enhance their chances of becoming rich.

This section has reviewed the question of management capabilities from the more personal angle of the entrepreneur or the initial small entrepreneur team. The question of the management skills required from the overall company viewpoint and the way these change with its type and size is considered on p. 20.

■ THE ENTREPRENEUR AND THE GROWING COMPANY

One further point on the initial management team is how the management structure should change to effectively cope with company growth. The driving force behind a technology-based start up is usually the creativity of one person, the entrepreneur, and, as the business grows, it is still important to maintain an innovative staff and an environment which is conducive to further innovation.

In a small company it is not possible to shut the inventor in a box and tell him to get on with inventing. There has to be at the very least some interaction with the marketing, financial, personnel and trading aspects of the business. This is a good thing, since it forces the inventor, who may be an academic with no real business experience, to come into contact with commercial reality and to recognize that his skills, although important, are not the only ones necessary.

Once the company starts to grow, however, the time necessary to handle the routine business aspects can become rapidly overwhelming, and the inventive energy and creativity of the entrepreneur are bound to suffer. Some means must be found of allowing invention to continue or the company will fail by being a one-product firm.

There are two questions here. Firstly, should an innovator be diverted from his valuable strength of providing new ideas and creativity to spend most of his time on the day-to-day management of a company? Secondly, is such a person the best person for this or is he even *capable* of being such a manager?

The person with the creative flair is not usually the kind of person to pore over cash flow or project schedules. The answer soon becomes compelling: the inventor/entrepreneur, despite being the one with the idea and drive to get the company off the ground *must* give way to someone with more inclination towards, and experience of, general management. There are exceptions to this rule, but it does hold in general. There are many instances of such innovators that cannot be persuaded to step aside and the decision is then forced upon them by their boards, who, having broader commercial experience, can identify the need.

There is also, when the time comes to seek finance for expansion, the question of the attitude of investors. Whereas a perceptive, intelligent technologist who is willing to learn some management skills may be perfectly capable of running a company with a turnover of around £250,000 per annum with four or five others, it takes an entirely different kind of person to manage an expanded firm employing 40 or 50 people with an annual turnover of several million pounds.

The hard fact is that running a larger firm is not the same as running a small one. This is why some businessmen take the decision, conscious or otherwise, not to expand. This may be an acceptable attitude for a builder and decorator, but not for a technology-based business which has but a short time to develop its market before the competition arrives.

There is no reason, of course, why the inventor should step *down*. Many accept an alternative role which suits them far better in a growing company so that they can continue to innovate, devise new products or enhance old ones. They can do this from either inside the company or, for instance, back in the university department. Here, with the university's consent, they can retain the academic contacts and use of equipment to continue the search for new and improved products.

For instance, if the originator of the business is offered a seat on the Board of a now refinanced and dynamically growing company – perhaps as Technical or Research Director – with suitable remuneration, this might prove adequate compensation for losing overall control. Indeed, it might, on reflection, be more than adequate, since it frees him from the day-to-day problems of running an increasingly demanding operation, to concentrate on what he is good at and perhaps what he really wants to do, viz, to develop new products.

The entrepreneur, faced with this decision, will first have to consider whether the solution offered is what is needed and whether the financial terms, etc., are personally acceptable. If they are, he must then ask whether having 60% (say) of something is better than having 100% of nothing.

☐ Dr. John Slack of Aston Molecules, based at Aston Science Park in Birmingham, is typical of an entrepreneur who has successfully faced these challenges, and I am indebted to him for the following brief history. His is typical of the kind of innovative company that recognizes the need to take a totally businesslike approach.

John Slack is a talented pharmaceutical chemist who had worked at the University of Aston in the Cancer Research Campaign's Experimental Chemotherapy Unit. The Unit's work involved goal-oriented research, and it was attached to the University's Pharmacy Department, a multi-disciplinary department with experience of consultancy work. Undertaking a certain amount of contract work from his position within the University suggested to John and his colleagues that the formation of an independent company to undertake this highly-specialized work on behalf of major pharmaceutical companies would be a worthwhile business project.

The original company was located in the University; indeed, the University took an equity stake in it. Small contracts were obtained, which paid for running expenses and patenting costs. Things developed quite quickly, and in 1986 the company associated itself with Aston Science Park. 'The decision to move from my academic roots was a major one', John explains, 'and I took it for a whole mixture of reasons, partly personal and partly to give me and my other scientific colleagues a better view of the commercial reality of the project'.

'For us, it has been important to maintain close links with the University, and a proper agreement has been reached on the use of their expertise and of facilities which we don't have in-house. It has also been valuable to have the business advice and support from the team at Aston Science Park. I think this source of independent advice was the most critical part of the move, for me'.

'As the Company grew, it became apparent that the business skills needed to run it were not only becoming much broader, but also that I, personally, did not have those skills. Added to this, the time I could devote to the special expertise which I *did* have, of pharmaceutical chemistry, would become diluted. I decided that it would be better if I spent more time at the thing I was good at and which I enjoyed. So we brought in an experienced Chief Executive and I have the role of Technical and Marketing Director.'

Aston Molecules moved to larger premises within the Science Park in 1991, and now employs 14 people, whilst retaining a close relationship with the University. □

Let us take time out for a minute to think about the personal qualities needed for management. We thought about them in terms of the entrepreneur or 'champion' on p. 6, but we now need to address similar questions about the management as a whole. What *are* the qualities that make a good manager? Are the same qualities needed for a small consultancy and a manufacturing operation? Of course, different skills and experience are necessary, but are the *qualities* similar?

Many books have been written on this, frequently implying that the qualities needed for management can be easily learnt; they offer a wide range of theories and buzz-words which seem to offer the potential manager the same kind of choice and prospect of instant success as do popular slimming diets. [Look on any station or airport bookstall. The Cambridge Management System; The Harvard

Slimming Plan; The F-Plan Manager; The One Minute Diet. Can you tell which is true or false?]

The bottom line is:

(*a*) a good manager can be made significantly better by training and experience,
(*b*) a poor manager can seldom be turned into a good one, and
(*c*) there are really four qualities which are important:

Vision. The ability to see what 'might be' rather than be satisfied with 'what is'. This is not quite the same as having objectives: the objectives are the set targets which, when collectively achieved, realise the vision. Vision is a key attribute, and, coupled with determination, constitutes the inner drive of many successful businessmen.

The ability to communicate. A prophet or hermit can keep his vision to himself. A manager must be able to explain his vision, objectives and needs to his investors, customers and staff. Also, communication is two-way: he should have the ability to *listen*. This is much rarer than the ability to talk, but careful attention to and assessment of people's responses is very important to all decision making.

The ability to inspire. It is important to be able to communicate not only the vision, but to also be able to inspire and encourage staff and investors to share in it enthusiastically. This is not just a matter of having verbal ability. It is achieved by convincing staff how good things can be for them if they can play a part in achieving the vision, showing genuine concern for their welfare and praising them for their successes, however small. A loyal, committed staff is an *invaluable* asset, which a good manager will work quietly and constantly to maintain. Good managers inspire by example. If the staff know a manager is at the pub every lunchtime or on the golf course every other day, his or her words, however poetic, will carry little weight.

Perseverance. Sheer dogged persistence! The good manager has to have an innate cussedness that he or she will just not give up or be deterred by initial difficulties. In other words, he must have the determination to succeed.

Just how do *you* measure up? To be first-rate at all of these would make you either a paragon or a liar. If you are going to be a good manager you will need to have all these attributes to some extent. If you are honest, you will probably know your weaker points; these are the ones to work at, and to take pains to improve over.

Types of technology-based business

■ HARD AND SOFT COMPANIES

One can make the distinction between so-called 'hard' and 'soft' companies, a nomenclature which reflects the amount of investment and the risk involved.

A consultancy firm is a typical *soft* company. Little is usually required in terms of overheads besides an office and a telephone line, and sometimes even the office is not necessary. There is no need for knowledge of complex matters like production planning or stock control and the management skills necessary are minimal.

A *hard* company is one which is engaged in manufacture, marketing, distribution and sales. The investment is higher, the number of personnel at start-up is greater, and there is a level of management required which is well above that of the soft company.

It has often been suggested that there is a steady progression by some companies from soft to hard. If the initial consultancy company is successful and the managers gain in experience and knowledge of their market, so they may progress to contract R&D work, or perhaps to manufacturing components or sub-assemblies for specific clients. If this is also going well, their next stage is to progress to full-scale manufacturing and marketing for a broader market. (This sounds a satisfyingly logical theory and no doubt successful instances can be cited. However, a recent survey by Professor David Storey on behalf of the UK Science Parks Association did not find any evidence for this progression within Science Park companies.)

An examination of the distinction is useful because it reveals several differences between the risks involved in starting a consultancy and those involved in initiating a manufacturing venture.

■ CONSULTANCY

A substantial number of research scientists or academics start consultancy or technical advisory businesses, perhaps having made contact with potential clients through their reputation in research and as experts in their field. The market is thus already known in part to the consultant and vice versa.

A university will often make its equipment available for any necessary experimentation, and the cost of additional overheads can be kept low. There is no question of distribution, and marketing and advertising can be initiated at a minimum level, just enough to let your potential clients know you are in business. A consultancy is thus a low cost route to starting a business – but, of course, you have to be good in your field.

■ THE RESEARCH AND DEVELOPMENT COMPANY

While 'to be viable' is a common objective, businesses differ substantially, not only in size and in the markets they address, but also in the level and degree of technology involved.

At one extreme is the research and development company, whose task is not to sell products or services direct to the end-user, but to research and develop new products for others to manufacture, market and sell. Large corporations such as pharmaceutical firms often 'spin out' R&D groups into discrete companies with separate boards, budgets and locations. They are distanced from the main company in order to get them away from the day-to-day pressures of production and marketing, and also to clarify their financial and organizational position within the group.

The same principles can apply to an independent, entrepreneurial R&D company. For example, a group of academics or technicians, highly expert in their fields, might recognize that their skills could be put to good use in designing and developing new products. On the other hand they may have no wish to develop all the production and marketing infrastructure to make and sell in competition to the multinationals. Their best choice could be to form a company purely to support the R&D and take ownership of the intellectual property generated. Licences to manufacture and sell can then be offered to the larger firms with the appropriate facilities and market.

This can be a good strategy, but it has its drawbacks. Some outside investment will probably be needed for what sounds like an unusual and esoteric product, and the investors will have to be persuaded that the market is there, the management's technical capability is good and that they will get their money back with interest within a reasonable time. Time can be important here, since many products based on highly advanced technology need a long time to develop; and the R&D phase can rapidly eat up money. Good management and tight project control are essential.

Speculative R&D, where there is no initial client for the design or prototype, carries the greater risk. It is therefore preferable to undertake contract R&D where a specific customer is identified from the outset. Naturally, investors far prefer a customer to be lined up beforehand.

The risks in R&D are higher than in consultancy, but the potential rewards are greater. The type of product can range widely, from software, through the application and design of specialized high vacuum equipment to production of genetically engineered species. Frequently, an academic or research institution acts as the initial home of such companies, and the equipment and information facilities of the institution are available to the entrepreneurs, sometimes allowing ideas and designs to be considered at little cost to them.

■ MANUFACTURING COMPANIES

Setting up a 'hard' manufacturing company involves greater investment, higher management skills, a longer time and a greater risk; but it can ultimately produce a greater reward.

The investment is higher because finance for R&D and production facilities must be found, as well as that required to promote, distribute and sell the product. For every £1 it costs to do the R&D and develop the prototype it will cost around £10 to get it into production, develop distribution and markets and launch the product itself. The maximum funding needed (the bottom of the trough in the cumulative cash flow curve) will thus be much higher.

A higher level of management skill is involved simply because there are more aspects to manage than in a consultancy or pure R&D company. Production scheduling, quality assurance, marketing, distribution and sales management will all be needed.

The size of the investment and risk depends on the market to a significant extent. A small, highly-specialized 'niche' market carries a lower risk. The company would not have been formed if this market had not been previously identified, and the potential customers would almost certainly have been known to the entrepreneurs beforehand. Examples might include the manufacture of superconducting magnets for research or medical use, or the production of specialized software for estate agents.

Many well-known firms incorporate components from other manufacturers into their products. If you have a plan to make, say, a much improved hard disk drive for a computer and can get the large manufacturers to incorporate your drive in their product, then you have the makings of a successful business. What's more, by selling to a limited number of customers, you can significantly reduce marketing costs.

There are disadvantages too. Being restricted to a limited number of very large customers means they can play off the few firms which are competing against each other, to reduce the price they have to pay.

This is a problem for any firm which relies solely on a few large customers. In addition, the advent of a new, better, competing product can mean that the sales can drop suddenly and catastrophically. It is essential to be continually developing new and better products yourself, to beat the competition at their own game.

The biggest challenge of all is to aim at the mass market, especially the consumer market. Few technology-based firms go directly from start-up to full-scale production for the mass market in one go. The costs are high, ranging from around £0.5 million to £5 million. A period of product development usually lasting between one to two years must be followed by an extensive marketing and sales campaign, maybe overseas as well. At the same time production and distribution facilities must be set up and put into operation. The necessary management expertise is thus substantial.

If the product is in an expanding market, the competition will not stand still. There must be regular reassessment of plans in the light of expected and unexpected threats from competing companies.

Subcontracting

Many companies wishing to innovate but which are deterred by the high cost of establishing and managing an extensive production facility overcome the problem by subcontracting construction and perhaps assembly elsewhere. Many 'makes' of personal computers now on sale are constructed by a relatively small number of production companies. They are then rebadged and sometimes reboxed before sale.

This can be a good, and sometimes the only, way out of the dilemma. Nevertheless, it is not without its pitfalls. The first problem is quality: you must specify the requirements very tightly to ensure the standard of quality needed to satisfy your customers will be met. Then there is the need to be sure that the schedule can be maintained, so that you have the right number of products ready at the time you need them, and not six months later.

This potential loss of control is the major hazard in subcontracting. The product *you* thought of and the business *you* started now becomes substantially dependent on a company outside your direct authority. Bigger companies are able to place subcontracts for similar products or sub-assemblies with several subcontractors (this is called 'second sourcing') but it is not usually possible for a new technology-based company to do this.

Besides loss of control there is loss of profit, since the subcontractor will need to take his cut. If you are making a good profit on your sales then this is acceptable, but the advent of competition may eventually force you to lower your price. The use of subcontractors may then not be so attractive.

Despite all this, subcontracting has its real advantages. If it did not, there would not be the current plethora of specialist and successful subcontractors to the computing, telecommunications and electronics industries. Not only does subcontracting reduce the level of investment required to launch a product, it also eliminates the need for the management of its production. It enables a company which has a good track record and strength in R&D to produce a marketable product without losing its essential R&D character. And it enables such firms to capitalize on their R&D quickly by turning it into products, and thus enhance the rate of their expansion.

Identify your USP

All successful businesses have a Unique Selling Point, or 'USP'. They have some feature which makes them stand out above the crowd, and which draws the attention of the market.

The 'USP' need not necessarily be the incorporation of the latest technology. Indeed, although the product may have this, it may not help to sell it in the least (*see* Chapter 9 p. 201). In many instances the technology may be, in the eye of the purchaser, an incidental matter. You, on the other hand, may recognize that it is really the technology which enables the product to achieve its unique characteristics.

I am writing this on a small 'notebook' computer. I know it has the latest large scale integration chips and a 386sx processor which enables multi-tasking, etc. Despite my awareness of the advanced technology, that isn't why I bought it. I bought it because it is lighter *than anything comparable on the market. Its lightness is its USP, and although the technology helps it to achieve it, the average user is quite unaware of it.*

There are so many things which can go to constitute a USP, and many of them have little to do with the technology at all. For instance, it is always possible to sell quality and service. A reputation for these can go a lot further than being at the leading-edge of technology.

This is related to the desire to please your customers (*see* Chapter 9 p. 204). It's not how you see your product and your company but how they see it which matters. You can beat the competition, even large companies, if you build up a reputation for quality and service.

Your delighted customers will tell others, and the business can grow even in the face of formidable competition.

The essentials of the new business

The essentials for any successful business, whether technology-based or not, can be described as the 'three Ms': market, management and money. Luck can also play a part, as can the current economic climate, but I am neither a fortune teller nor a politician (can you always tell the difference?).

All of these key elements must be encapsulated in a Business Plan, which is discussed in detail in Chapter 4. This Chapter now considers these elements in turn. Please remember that I am trying to encourage you to relate what I'm saying to your business idea; not to pour cold water on it or to make it seem more difficult to start a business than it really is. I mean to encourage you and your potential partners to do some realistic assessment – about yourselves, your idea, the opportunities you must seize and the threats you must face up to.

■ THE MARKET

There is no point at all in starting a business based on a product or service that nobody wants to buy. This may sound blindingly obvious, but it is a common error amongst inventors. Invention is but the first step to innovation. The question is not how clever or unusual it is, but will it sell? In other words, is there a *market* for it?

There is a natural tendency for the technologist to be impressed by sheer technological achievement, and to feel that the quantum leap in this is a guarantee of future sales (*see* p. 201). This is by no means necessarily so: there are many investors who have been persuaded by the eminence of the inventors and the technological brilliance of the concept to put money into commercialization of scientific discoveries which are still searching for a market.

*One such company had licensed a very sophisticated parallel array processor from the British Technology Group (*see *Chapter 5 p. 155) which had been invented by a group at University College, London. The company felt there must be a plethora of applications for the device, and staged a number of seminars to promote it, inviting senior development personnel from 20-30 top British companies. The objective was to invite the manufacturers to dream up ways in which the processor could enhance their products. None of them did.*

Conversely, a person with an eye for the market and preferably specialised knowledge of it, may identify a product incorporating only a moderate level of technology, which nevertheless has a good sales potential.

One client entrepreneur who came to my Newtech Innovation Centre was the manager of a small metal-forming company whose major product was a waste bin which was located on streets or exhibition sites. It was frequently purchased by local authorities or fast food shops, and carried advertising panels on it, exhorting shoppers and customers to keep the locality tidy by using the bin. Hardly a technological product, I thought, so I asked him how we might help.

'I don't know a thing about technology,' he said, 'but I do know the market for these bins, and if you could make me a rubbish bin that talked I know I could make a lot of money.'

So we made him his prototype talking bin and our computer experts put in the latest in speech synthesiser chips and high-efficiency solar cells to power it. It soon completed its trials, and was on sale: since then, they have been sold throughout the country and abroad. A success for the entrepreneur!

The point is that, despite the avowed involvement of technology, it involved applying already existing technology rather than breaking new ground. Although the technical experts in Newtech know quite a bit, they did not know the market for rubbish bins, which was the key factor ensuring the success of the invention.

It may be that you think there is a market and are right. For instance, perhaps you have been involved as an academic or a technician in a specific field of technology for some time, you have realized a market deficiency and are planning to fill the gap with your product or service. For instance, you may have been working on specialized signal amplifiers, high-resolution mass spectrometry or NMR, and have become so good at servicing or modifying the instruments in your care that you believe you could start up a business to do just this. You already know the competition, if any, and also all the other major users of such highly specialized equipment in Europe. They may have already asked for your advice and help in the past.

Nevertheless, you could also think there is a market and be wrong. Maybe your friends and colleagues have encouraged you when you asked them if they would buy your product or service if you decided to 'go it alone'; it would not be natural for them to say no and to discourage a colleague who was contemplating such a step. However, when your business is set up and you go to them and quote a price which to you seems quite realistic, they may then rapidly change their minds.

In other words, not only must there be a gap in the market, but there must be good evidence of a suitable gap between the price the market is prepared to pay and the cost price.

What real *quantitative evidence* have you? Supplement the knowledge of the potential market which you have acquired by experience with lots of other questions. 'If I started up on my own, would you buy my product or service? Would you be willing to pay (so much) for it? If not that much, then how much? How frequently would you buy? Have you heard if any one else is going to offer such a service? If you won't buy from me, why not? Is it price, quality, or what?'

During this investigative phase you may learn some uncomfortable facts. It is an important part of the process however, if you are to have a chance of satisfying, and, ideally *delighting* the market you are attempting to serve (*see* Chapter 9 p. 204).

A consultancy, research and development business or one which is based on a known 'niche' market usually starts because the entrepreneur gets to know or appreciate the existence of the market by personal or inside knowledge over a long period of time. Indeed, many technical consultants begin by doing a few jobs for their contacts.

A technology-based manufacturing business, however, should not start like this. The greater initial investment means that the market must be identified, quantified and surveyed before commencement. Indeed, the Business Plan will not get beyond the first hurdle unless it has (*see* Chapter 4 p. 95).

It is advisable for the market assessment to be independent of the management, or at least of the 'product champion'. Sometimes a more conservative estimate of the product's sales will result, which may not forecast disaster but will allow initial production to be more realistically estimated. Sometimes an independent assessment can reveal an additional, unexpected market.

Since most small manufacturing start-ups do not have a strategic marketing expert on the management, it could be advisable to employ a consultant, perhaps under the Enterprise Consultancy Initiative (Chapter 5 p. 122).

■ THE MANAGEMENT

Earlier, we considered management from the more personal viewpoint of the entrepreneur or 'product champion'. It is also important to view the management question from the standpoint of the company as a whole.

The quality of management is even more important than the identification of a substantial market, since a first-rate manager can sell more in a difficult market than a mediocre manager can to a good market. By far the largest cause of failures of small and medium-sized companies is the standard of management.

It is easy to fall into the trap, writing a book like this, to talk about 'the management' as though it is some abstract concept or commodity. The management, of course, is a collection of human beings, each with their own strengths and frailties. The essential thing is that their strengths should be complementary and sufficiently wide to encompass all major aspects of running the business.

The personal characteristics of a good manager (*see* p. 11) are essentially the same whatever the size and growth rate of the company, but the required skills will be different.

Growth rate is not the same as size. Whereas the specialized consultancy may grow only slowly, the more successful technology-based businesses can grow very fast indeed, a feature which can bring considerable challenges, not to mention strains, to the management.

As an example, the sales of Microsoft, the US company responsible for the MS-DOS operating system which was first adopted by IBM and is now used by 70 million personal computers worldwide, grew from $8 million to $100 million in four years – a phenomemal rate of growth.

Interestingly, Bill Gates, Microsoft's Founder, Chairman and Chief Executive, almost bungled it. When IBM originally asked him to produce an operating system he had misgivings over whether his small and inexperienced company could cope. He recommended a competitor to IBM, but they could not agree on a deal and the request arrived back on his desk. From then on he has never looked back and has become one of the richest men in the US. □

Let us now look at the various types of technology-based businesses, assess their different management needs and the skills required to meet them.

The consultancy company

A small technology-based consultancy company may well consist of just one person at the outset. He or she will constitute the total human resources of the company and will thus have to perform the duties of the marketing manager, the sales manager, the financial manager and the administration manager, besides actually carrying out the consultancy itself. This is not impossible for a while, but any such company may soon find that the joint pressures of administration and financial control *and* doing a first class consultancy job for the customer are too much.

The solution at this stage will probably not be to increase the management staff, but to engage an administrator or personal assistant to offset some of the non-specialist load. Until the turnover increases substantially, part-time accountancy and financial management input will probably be sufficient. Up to a point, the higher daily rates for such temporary staff will be more than offset by the saving on the cost of engaging such a person full-time. When that point is reached, the decision to increase the staff complement will have to be faced. Remember, though, that administrative and financial staff are not 'productive': they do not directly produce income from their labour. Only adding to the complement of consultant staff can do this.

Eventually, it may be necessary to engage additional staff to undertake marketing. Although the consultants themselves will still be the best people to actually sell their own skills, there will come a point where the marketing is better done by someone with experience in promotion and market strategy.

If you are a skilled technical consultant, do not underestimate your ability to sell, and more to the point, do not let suave or bombastic marketing people tell you that you cannot sell. The customer can be encouraged by good marketing, but when it comes to the actual sale, it is the technical expert who scores every time: the customer can easily detect a sales or marketing man who does not fully understand the product, whereas the depth of knowledge of the technologist himself is immediately apparent.

Research and production companies

Both R&D companies and production companies require a broader range of management skills than a consultancy. A well-balanced management team will have the right mix of these. The blending of two entrepreneurs, one with scientific or technical skills and the other with commercial or financial experience, can make a formidable and successful team. On the other hand, two brilliant scientists can produce a disaster.

This should not be too difficult to believe, but if there are doubts, ask the banks or the venture capitalists, the people who risk their (and other people's) money by investing in new businesses. They invariably prefer to back a management team with a blend of complementary skills. A management team consisting solely of scientists or only of marketing experts is like a football team with only defenders (I've seen a few of those, too!). They're not worth supporting.

On reading this, it may possibly be dawning on a few that they or their team lack the range of skills necessary for success. If so, do not be discouraged: to recognize shortcomings is a major step towards success. The task now is to remedy the defect.

One source of commercially-experienced people is the management staff shed by large organizations undergoing what is euphemistically known as 'rationalization'. There has been an increase in the availability of such people recently, from major industry, banks and large consultancy organizations. They still have great energy, often have aspirations of starting a company themselves and can be a tremendous benefit to any small company seeking their management experience.

Executive Recruitment Agencies, Enterprise Agencies, Science Parks and Innovation Centres often know such people or have them on their books, and will be pleased to help.

Another source is the small investor with business experience, who not only wants to invest in good business ideas but also enjoys assisting in the management. They have the advantage of coming with money attached, but it is important that they understand and appreciate the nature of technology-based business. They must be prepared to wait for long R&D phases and for a market to be established before they begin to see a return on their money.

However you cope with it, there is inherently a much broader range of skills required to manage an R&D or production company than many other types of small businesses. In addition, the emphasis in R&D management is different to production: in one, the stress is on project control whereas, in the other, it is more on optimization of production efficiency and sales.

The *finance* involved is also much greater and it needs more careful control – especially of R&D costs, which have a tendency to overrun. The larger sums necessitate the establishment of systems to monitor cash flow, purchases, stock, work-in-progress and collection of debts (*see* Chapter 7). A financial director or manager is thus a necessary appointment.

Production control is a management expertise in itself, involving both interpersonal skills and planning know-how. Quality must not be compromised but the schedule of production must be met. The alternative of subcontracting may, however, be worth consideration (*see* p. 16).

Then there is marketing, distribution, sales and after-sales service. A 'pure' R&D company will perhaps not need all of these if it is designing prototypes for specific clients, but a production company will. Marketing is an important skill, and it is unusual for a technically-minded entrepreneur to have such skills or experience.

In addition, the market for many technology-based businesses is international (indeed, some products are so sophisticated that the firm can only be viable if it addresses an international market), which will require additional, expensive but necessary management expertise.

Even if all these extra management skills are available, there are still consequent dangers. For instance, the innovative company which has successfully lived on its creativity is in danger of losing its essential 'sparkle' if it becomes diverted to the problems of gearing up to production and then managing it. The day-to-day problems of operation can totally stifle a management which were once proud of their creativity and innovation.

There are also the tensions which can result from the conflicting priorities of the technical, production, marketing and financial departments.

The Marketing people can be screaming at Production for the new product which they know they can sell 'but only if it's ready tomorrow'; the Production Manager is at odds with the Technical Manager for not producing the drawing for the vital modification needed until a fortnight after the deadline for commencement of production; the Technical Manager has made an enemy of the Financial Manager by over-running his budget again; and the Financial Manager has been up half the night trying to make the cash flow work out.

The Managing Director, who once was half of a cooperative and creative team with the person who is now Technical Manager, wonders what's hit him. Only good general management can keep a team working together, and any manager of a manufacturing company reading the above scenario will attest to its realism.

This is clearly a very different reality from the start-up, consultancy or small R&D company, and it was the reason for stressing earlier (*see* p. 9) that the entrepreneur who starts a business may *not* be the best person to run it when it has grown beyond a certain size.

The skills must fit the need

Most managers find themselves comfortable managing a firm of a fairly distinct size and growth rate. The typical entrepreneur is first class at creating a new company, injecting enthusiasm and creativity, finding financial backers and launching a new product onto the market. Such a manager will not feel happy if he is put in charge of a larger production company, where he is distanced from the technical or marketing work he has been used to.

Equally, some managers are most at home in a rapidly growing company. They like the challenge of growth, which involves the penetration of new markets and the introduction of a more systematic management into the organization. Others are most competent at running a larger company with a range of managers and production already in place. They are the kind of people who can handle – preferably by circumvention – the kind of management in-fighting implied above.

In summary, a rapidly growing company requires different management skills from either the small one, or a large but constant-sized one. A firm employing five people can succeed with a much less-structured management style than one with 50. As a company grows, its internal structure will have to change, and rapid growth may demand a succession of such changes in a short time. Each will have to be carefully managed. Each phase of growth will bring the need for revised systems of financial and administrative control.

■ THE MONEY

The raising of finance for and financial control of businesses are so important that they are given Chapters to themselves. Nevertheless, there are some general points which can conveniently be stated here.

Firstly, raising finance of any size can be very time-consuming. The preparation of a Business Plan (also given its own Chapter) can take some time, and the professional advice which is usually necessary is not cheap.

Secondly, you should be prepared to put up some of the start-up money necessary yourself. How much will depend on your personal circumstances and the needs of the project, but no one is likely to back your idea unless you yourself are prepared to risk some of your own money on it.

It may seem obvious, but when I am assessing business plans from clients of my Innovation Centre, this is one of the criteria I find exceedingly useful. Before I pass on a plan to the banks or other sources of investment, I ask

(a) *how much is the client putting in himself, and*
(b) *in principle, would* I *be prepared to invest in it. If I have any doubts, I then try to analyse why.*

Thirdly, don't underestimate or try to minimise the amount needed. Of course, lavishness is to be avoided, but placing undue limitations

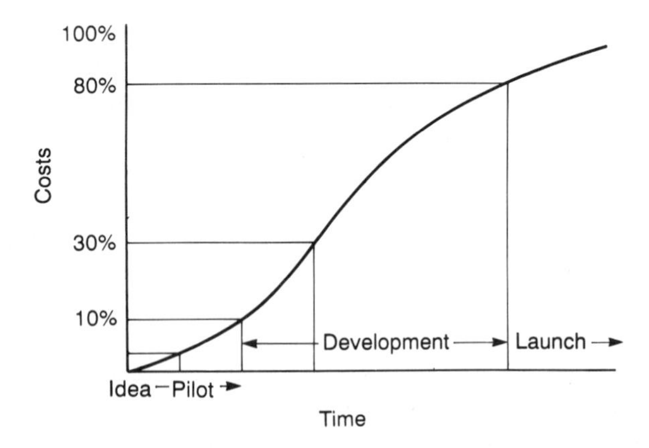

Fig. 1.1 Costs of each phase of launching a product

on the amount of finance can have the effect of limiting the amount for product development, delaying a product launch or attacking a narrower market. If you are planning to develop an idea into a reliable, high-quality finished product, do not underestimate the costs of development and preparation for production.

Fig. 1.1 illustrates this. If you are a technologist who has just spent a year and what to you is a substantial sum of money on the first working prototype, you have probably only gone a little way along the road. It will probably take another 1-2 years and 3-10 times the amount of money to get it into production.

Fourthly, don't forget the need for working capital. There is a tendency for lending organizations to be more concerned with covering the costs of capital equipment and ignoring the money needed to cover the costs of trading until the sales income actually comes in. This is the working capital (*see* Chapter 7 p. 176).

Finally, the need for a good Business Plan is paramount. *All* the aspects dealt with here, the product or service, the market, management skills and financial requirements should be incorporated in the Plan.

It had better be good. Venture capital for technology-based businesses, especially start-ups, has never been easy to obtain, and the banks are now tightening up on their lending, having been rather too profligate in lending to small firms in latter years, many of which folded in the recession of the turn of the decade. It is your Business Plan which is your initial introduction to the world of bank finance and venture capital, and it is thus probably the key to unlock your future.

Take advice

The main message of this Chapter is that there are a lot of things to think about when starting a technology-based business. Even the launch of a small consultancy or technical service firm involves consideration of a range of complex matters, many of which will be new ground to the technically-minded entrepreneur.

Maybe you have read through the Chapter and feel somewhat daunted, or even deterred. This does not matter, provided you feel more aware of the importance of a particular aspect or somewhat clearer about some feature of starting a business. It is better to be cautious but informed than bold and ignorant!

Do not be totally put off unless you now realize that you were totally on the wrong track. The thing to do now is to reassess the project, seeking independent, reliable and experienced advice where appropriate.

There is an extensive business advice network in the UK and the Appendix gives details of where such centres are and how to get in touch. In fact, there are now probably too many advice centres: Training and Enterprise Councils, Enterprise Agencies, Small Business Advice Centres (some run by local Government, some by Development Agencies and others too), Business and Innovation Centres and Science Parks. There is a need for rationalization, or at least an attempt to avoid wholesale duplication of services.

Nevertheless, such an advice centre is the place to start, ensuring first that the staff are accustomed to understanding and dealing with *technology-based* projects. They will all tell you that they are, but frequently this means that they helped someone to do something like start up a satellite television installation and repair service. Ask them how many projects they've helped in the last year with a significant R&D content, or how many technical consultants they have as clients, and in what fields.

This is important, since, as this book indicates, there *are* problems and opportunities specific to such projects, and a centre which is more accustomed to advising on retail or light manufacturing start ups may not give the best advice. The typical high street bank manager is usually a disaster when it comes to talking about technology, although the banks are now realizing this and are attempting to do something about it (*see* below).

The local office of the Department of Trade and Industry may be a good place to start, if you are unsure. Alternatively, try the local Business and Innovation Centre or Science Park. The National Westminster Bank has a Technology Group, with Technology Managers strategically placed throughout the country; this is a substantial commitment, and should be increasingly relevant to the entrepreneur involved with technology. Barclays Bank also has a High Technology Team, with a record of assisting technology-based business (*see* p. 133 gives more comprehensive details.)

All business involves risk, and the objective of good advice is to minimize that risk. I have seen business projects benefit from thousands of pounds of grants for which the proposers did not know they were eligible; entrepreneurs made aware of the existence of unexpected hazards and unexpected opportunities, such as new markets; technical problems overcome by the input of specialized consultancy; new management introduced to firms that desperately needed revitalization. All of this was due to taking timely advice.

Chapter 2
TECHNOLOGY AND TODAY'S SOCIETY

Introduction

This Chapter will attempt to outline the currently more important growth areas of technology. It is a hazardous business, to try to predict the areas of technology which will be exploited as we move into the 21st Century. Indeed, if I knew precisely and certainly where the real growth areas were, I would be investing heavily rather than divulging the information to you!

Nevertheless, there are some firm pointers for the future, and this section is designed to be of use not just to the scientist, who will know a fair amount of it anyway. The investor or businessman who is not a scientist, but who may be considering applying his money or skills to this important area of enterprise, may find some general explanations and guidelines of value.

One complexity is that the growth areas of technology are not in watertight categories, but overlap and influence one another. The application of control engineering and computing has been vital to the development of continuous fermentation. The areas of computing, electronics and robotics converge one into another. So the classification is one for our convenience.

A technological society

We are a society permeated by technology. New technology, high technology, 'leading-edge' technology, 'state-of-the-art' technology. We are all familiar with these phrases which are often used by people with little familiarity with science or technology. When my Innovation Centre was opened in 1985, BBC Wales described it as a 'Renovation Centre' and the most technological question the Press asked was 'How many computers will it have?'.

The problem seems to be that technology has become a major force in most people's lives whether they understand it or not – and too many don't. Can we do better than this, and try to focus on what most of us mean when we talk about the role of technology in today's society? Perhaps the pivotal role of technology in modern industry will then be more appreciated.

There is nothing inherently wrong in being impressed by or even delighted with technology for its own sake. In August 1988, the tiny space craft Voyager 2 sent back pictures of Neptune and its moon, Triton, the largest in the Solar System. Being the so-called 'silly season', this was front page news at the time, and, besides the wonder of the pictures themselves the Press did manage to convey a sense of awe over the technology involved in getting them back to Earth. Of a puny 20 watt transmitter sending signals which took four hours to get here at the speed of light. Of the detection and amplification of the signal which reached us at a fraction of a billionth of a watt. Of the ability of the technicians to guide it to within 30 kilometres of its course over the 12-year voyage, and to precisely rotate the craft at a rate one-tenth that of the hour-hand of a clock.

This example is about as neutral an instance of technological application as I can think of. Indeed, I would argue that technology itself is as neutral as a knife. A knife can be used to kill someone, or, in a surgical operation, to save a life. The technology employed in Voyager 2 was developed and used in the 1991 Gulf War to pinpoint targets and destroy them; whether or not the war or these methods were justified is not the subject of this book, but it illustrates the point. It is what we *do* with the technology that matters.

The examples given so far are fairly spectacular ones. To really appreciate the way in which technology has become an integral part of life, often without us being overtly aware of it, we need a more everyday example.

Consider a lecture visit I have just made to Germany. The first communication was a *fax* – unheard of a few years ago. I confirmed by an ISTD phone call, which involved fibre optics and digital exchanges. My air tickets were booked via a complex computer system, and when I checked in all details were remotely compared and checked. My baggage was weighed electronically, and its weight was added to give a running total of the cargo load. It was security checked by a low dose X-ray machine.

This does not include the aircraft itself, with its new materials technology, its airframe and engine design. And it does not include the computer I used to write my talk and generate my slides.

The sum of all this is that we often estimate the degree of development of a nation not just by its Gross National Product or its number of Symphony Orchestras (pity!), but by its degree of technical development. Essentially this means the *exploitation* of technology, and it is this aspect which is the subject of this book.

The flow of technology

There are several ways in which technology may be subdivided. In terms of its application, the one used by Douglas McQueen of Chalmers Innovation Centre in Sweden seems the most operationally useful and enables us to appreciate the flow of technology between academe and industry. This is illustrated by Fig. 2.1. He distinguishes between:

Basic Technology. Basic Technology is the core of most undergraduate courses, and includes Mathematics, Chemistry, Physics and Biology.

Key Technology. Denotes those subjects most relevant to current industry, such as Informatics, Biotechnology, Materials Technology, CAD/CAM, Robotics, etc.

Pacemaker Technology. Those areas which are likely to be of great

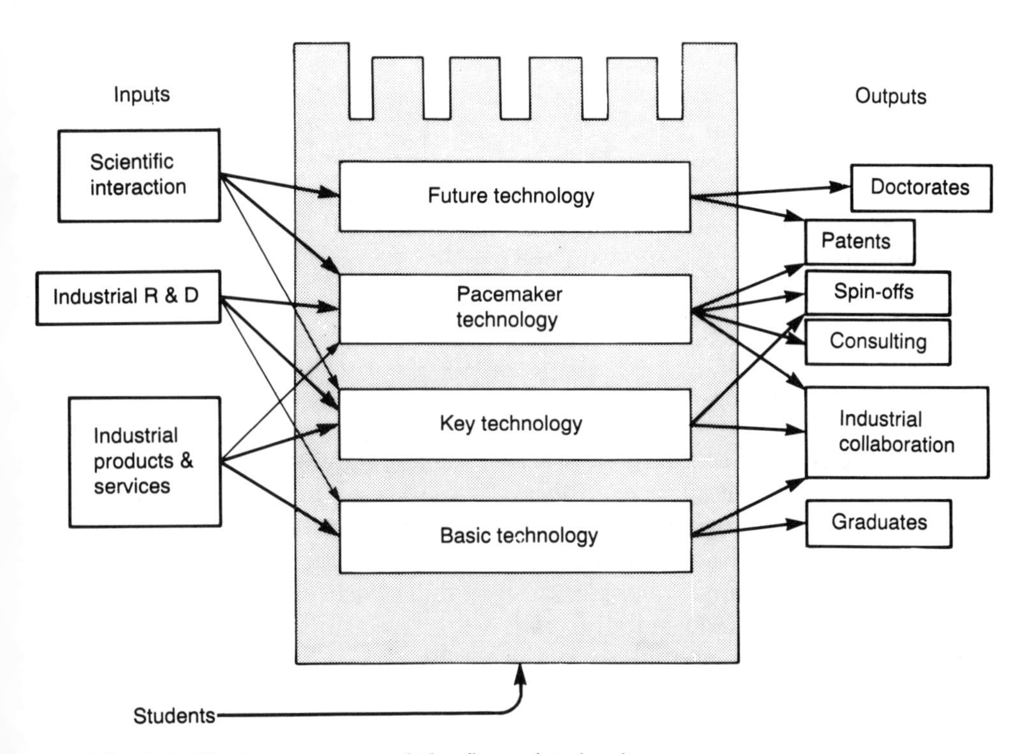

Fig. 2.1 The ivory tower and the flow of technology
(with kind permission from D. H. McQueen)

significance in the near future but which are not yet the subject of widespread application, such as Artificial Intelligence, Sub-micron Electronics and Genetic Engineering; and
Future Technology. Future Technology includes Photonics, Neuro-informatics, Ambient Temperature Superconductivity and Bio-electronics. These are some areas which are now the subject of research and speculation, but which may hold a bright commercial future. It would be nice to be able to identify which will be successful in advance.

The dividing line between these categories is somewhat fuzzy, and, as a particular technology becomes the principle behind a number of commercial applications, so it can move up the list.

Using these categorizations, one can more readily appreciate the way in which the academic/industrial interface operates. The Basic Technologies are dominated by large industries, so that entry into such a market by a smaller firm is difficult. Companies formed by direct spin off from a university are usually based on Key or Pace-maker Technologies, since, although the potential market for Future Technologies may be apparent, the products and customers are not.

Douglas McQueen also emphasises the two-way flow of technology between industry and universities or centres of research, which those of us involved in running Science Parks are well aware of. The company with close links to an academic department, whether or not the principal founder came from the department or not, will be feeding ideas back to the department, thus providing a close operational link between the university and the industrial world. Hence the enhancement of the university's role in industry and the community which Science Parks and academic 'spin offs' can bring.

The exploitation of technology by the smaller firm

The exploitation of totally new areas of science and technology is not usually for the small firm. It normally requires enormous resources, both technical and financial, and several years of research and development. Projects of this kind involve a strong and experienced management team, novel and protected intellectual property, major investment in R&D and the patience of investors to wait several years for a return on their capital. It can be done, but today's attitude of 'short-termism' amongst UK financial institutions does not assist.

Smaller firms and new start-ups can still play a vital place in the exploitation of Key Technology. Substantial development may still be needed, but the time taken between start up and product launch is usually shorter.

Often, the smaller firm is better able to exploit technology than the larger one. This is because:

- The smaller firm can develop it cheaper or quicker than a large one, because the new product may not be a priority for the larger firm.
- It may be better at identifying and exploiting a market niche. This may be in a market which the larger company, preoccupied with its mainstream product, has not identified, or, if it has, it has been given low priority because it does not fit well into its corporate strategy or current product portfolio.
- The man who sees an idea for a new product can usually satisfy his aspirations better in a small company than as a minor part of a large one. A product 'champion' running a small firm has all the motivation in the world to make it successful. In a larger firm, these can get changed, frustrated or swallowed up in corporate decisions.
- There is a wealth of opportunities, in, for instance, peripherals and servicing for the major projects of larger firms, which the smaller company can exploit. Valves for bio-reactors and new accessories for computers, for instance – many of the markets for such products have been generated by major technology-based projects. This is true not only of domestic and industrial markets, but of the defence industry also. Such opportunities are often recognized first by potential entrepreneurs working in larger companies, who decide that there is a market which they themselves can best fill. They then move out to form their own small firms, becoming the 'product champion'.

Growth areas of technology

The rest of this Chapter is devoted to a brief review of the major technologies, focusing on the more likely growth areas. It is necessarily brief and somewhat speculative, but the enormous potential for commercial exploitation should be apparent.

The very nature of technological development and innovation teaches that new developments often come from quite unexpected areas, frequently at the interface between two disciplines or technologies. It is also worth noting that some of the most successful innovations come not from totally new developments in technology, but from the novel application of existing modern technology.

Materials technology

The advent of new materials has influenced our lives as much as computers and instant communications, but they are perhaps neither so obvious nor so glamorous. New materials are not advertised in the business sections of Sunday papers each week, but they play a part in our daily lives.

Of course, there are the headline-catching developments such as the use of carbon fibre composites for aircraft wings or new composite materials for surgical prostheses. But there are also better materials to make lamp posts from, and the potential market for these is just as large.

Current developments in materials could be classified as either the enhancement of their mechanical properties or of their optical and electronic properties.

■ CERAMICS

Materials capable of withstanding extremes of temperature yet having sufficient mechanical strength are important in many circumstances. Ceramics have found recent application in protecting the nose of space shuttles against frictional heating during re-entry into the Earth's atmosphere. More prosaic applications of such refractory materials are in lining of furnaces.

Developments in compacting ceramics have provided new opportunities. For instance, some of the many tonnes of 'fly-ash' rejected and dumped from power stations each week is now being compressed into building blocks for internal walls.

■ COMPOSITE MATERIALS

New composite materials are having a relatively unpublicized but significant influence. They include fibre-reinforced plastics for construction, and are used in a myriad of ways, frequently displacing metals to give strong, attractive and corrosion free products. The glass fibres used to reinforce some plastics are simpler to produce than telecommunications fibres (*see* p. 35); there is no need to make them with two concentric layers, since their optical properties are not important.

Composites are frequently lighter than the metals they replace, so are ideal in situations, ranging from components of spacecraft to racing bicycles, where the saving of weight is at a premium. The development and production of these materials has taken place within the larger chemical companies, but applications of their uses continue to be invented by smaller firms.

It is worth noting that one of man's oldest materials, wood, is undergoing something of a renaissance. Methods of compacting and bonding wood have produced composites of remarkable strength, fire-resistance or cost-effectiveness. These are used both in construction and furniture.

■ PLASTICS

Sometimes it is just as important for materials *not* to be too durable, and the developments in biodegradable plastics promise to have a positive effect on the environment.

ICI's development of polyhydroxbutyrate (PHB), with the trademark of Biopol has now reached the production stage, and the objective is to be making 5,000 tonnes a year in the mid-1990s. This material is made by bacteria from glucose, and, once its use is complete, soil bacteria degrade it back to carbon dioxide.

A material with comparable properties but made from starch is also being developed by Feruzzi in Italy. Neither of these plastics are cheap (there is a seven-fold price differential over polythene in finished products), and there is still concern over the environmental benefits of biodegradability *versus* recycling.

Other plastics have also been developed and produced which are photodegradable, so that they lose their strength and eventually disintegrate when exposed to sunlight.

As with other advanced materials, the smaller firm does not have the resources to research and develop the material in the first place, but they can secure good markets for themselves by being aware of the enhanced properties of the new materials and developing specific products which take advantage of these.

■ OPTICAL FIBRES

Bridging the gap between materials whose benefits are physical and those whose benefits are optical and electronic, are optical fibres. There is an enormous demand for low-loss optical fibres for telecommunications, since the information-carrying capacity of optical fibres is much greater than that of metal conductors.

They operate by trapping light within them by total internal reflection, and it is important that the refractive index at their core is around 1% higher than the peripheral cladding. Their properties can be further modified by 'doping' with specific contaminants, and by surface coating. Their first use was in medicine, where fibre-optic light guides have enabled doctors to peer inside various parts of the body with a minimum of invasion.

While obviously very important, this use has been overshadowed in commercial terms by their communications application. Optical communications fibres are made of highly transparent special glass, and are between 0.123 and 0.5 mm in diameter. A beam of light, modulated (i.e. varied) so as to carry information, can be transmitted at the speed of light along the fibre, and the only losses of intensity in such a pure glass come from the very small amount of scattering and absorption; the integrity of the information is retained after passage through the fibre.

The fibres can be joined, welded or glued so that their cores match. Present-day fibres can send signals for some 50-60 kilometres before amplification is needed. Zero-dispersion fibres, working at a particular wavelength of light, can be used for more long distances, such as submarine cables.

Besides the intensive interest in the materials aspect, substantial research is in progress in methods of amplification and switching (i.e. directing) signals carried by optical fibres (*see* p. 44).

The optical (or, strictly, the electromagnetic) properties of fibres continues to excite interest. Joint work between US and Russian research groups have developed a means of guiding and focusing X-rays by using bundles of fibres. Walter Gibson at the State University of New York and Muradin Kumakhov of the Institute of Atomic Energy in Moscow have used very fine hollow glass capillary tubes to achieve focused X-rays with very high intensities.

This invention should result in the possibility of producing microchips by lithography (i.e. directing radiation onto silicon wafers through holes cut in a mask) using X-rays from conventional sources. The use of X-rays enable much greater resolution than visible or ultra violet light, and thus enable more features to be packed on to a given area of silicon.

Other applications in medicine are also possible, and a company has been formed to exploit the invention □.

■ NEW SUPERCONDUCTORS

Innovation is often said to proceed systematically from theory to application. This is a myth. The application of steam energy to driving machinery, a principal development of the Industrial Revolution, was not discovered by consideration of the theory. On the contrary, the science of thermodynamics arose out of observation of the practical application.

So it is with the new 'high temperature' superconductors, the discovery of which led to one of the fastest Nobel Prizes in history for Bednorz and Muller, in 1987. Before their discovery, it was assumed from theory that the highest 'critical temperature' (above which the material would show ordinary conductivity) was around 25-30°K. This temperature, only just above absolute zero, meant that superconducting wires and coils had to be cooled in liquid helium. This is expensive, but feasible for some applications.

In 1985, Bednorz and Muller, working in IBM's Zürich Laboratories, serendipitously discovered a composite material which had a critical temperature of 30-40°K, and subsequent work by them and other groups has led to materials which are superconducting at liquid nitrogen temperatures – a much more practical proposition for applications. At the time of writing the highest critical temperature published is 125°K.

There has been an explosion of research and development, and a cascade of patent applications on new 'higher temperature' superconductors, but so far little practical application of the invention. The problem is that the new materials are not particularly malleable or ductile, so it is difficult to fabricate wires and coils from them.

No doubt, however, this problem will soon be solved, since the new materials offer a tremendous potential for a range of industrial applications, including electricity generation and transmission, and anything which requires a high magnetic field.

Returning to the point made at the start of this section, the discovery of these new materials has led to much head-scratching amongst the theoreticians, and the theory of superconductivity will have to be revised or modified to deal with the new phenomenon. At the time of writing, no clear theory has emerged.

■ SEMICONDUCTORS

Research on the materials aspects of semiconductors continues apace. Silicon is being replaced by gallium arsenide for some applications, but the bulk of the work is on dopants and surface coatings, and the methods used to etch or imprint the circuitry on them. The overall objectives are speed and the connected ability to be able to pack more and more elements into the chip.

The commercial pay-off is in supporting more and more computing power in less and less space. Research into the properties of semiconductors and their modification has underpinned the information technology which as a whole was the basis for the so-called 'Second Industrial Revolution'. Most developments have been undertaken by larger firms or by companies set up with public sector support. Research continues in several university departments, and the apparatus needed is very expensive. Smaller firms have on occasion developed innovations of this ancillary equipment.

Computing

At the heart of the revolution in information technology lies the quite fantastic development that has occurred in computer technology over the past 25-30 years, and which shows no sign of abating. Even if one ignores the totally impractical early machines based on thermionic valves, the increases in power and decreases in size have ranged over several orders of magnitude.

I can remember feeling very privileged during my postgraduate research in the early 70s to have access to a state-of-the-art computer which more or less filled a 150 sq.ft. office. This enabled me to get through statistical analyses in a fraction of the time needed by hand calculators (it took most of a morning with those old things to plot just one regression curve).

Yet the power of the '386 notebook' computer on which I am typing this manuscript far exceeds that of the roomful of transistors and enormous discs. It weighs less than three kilos, runs on batteries and fits comfortably in my briefcase.

One can usefully separate out developments into hardware and software. Of course, the computers themselves will continue to be developed. There is not only a drive for an increase in sheer speed, as in the Cray Supercomputers, but a parallel – and, some would say, a more practical – development in the 'transputer' or parallel-processor computer. These allow millions of calculations to be undertaken at one time, instead of sequentially, as in the ordinary computer.

Although one might imagine at first sight that such developments must demand the financial and research muscle of a large corporation, the fact is that many smaller groups, including commercial 'spin offs' from university departments, have been successful in this area.

Frequently, the small R&D group, having proved its concept, is bought out by the larger corporation for production and sales. However, this is a perfectly natural route for a smaller firm to take.

■ COMPUTER APPLICATIONS

The ubiquity of the microcomputer has led to the development of a multitude of peripherals and applications. This has been fertile ground for the computer-minded entrepreneur, and many have proved very successful commercially, starting as small businesses and sometimes growing to become market-leaders. However, many have fallen by the wayside, and these are the ones we *don't* hear about.

Some computing entrepreneurs have exploited the areas of computer add-ons such as additional memory, CD-ROMs (large read-only memories in a compact disc format), graphics and screen-enhancing systems, and hardware which enables a computer to communicate. These include things such as modems, networking systems and plug-in cards to enable the computer to act as a fax.

Then there are the peripherals such as printers, plotters, scanners and other image recognition devices. Innovative design and reliable products based on good perception of market need are vital to success, as in the software area (*see* p. 40).

Some small companies have successfully exploited the 'whole application system' approach. Rather than simply develop their own peripherals, they have concentrated on the specific application itself. There are innumerable examples, including security systems, electronic management of buildings (embracing security, heating and lighting, etc., and sometimes monitored totally remotely from the site itself), traffic control, fuel optimization for shipping, dimensional inspection of manufactured goods by optical measurement and comparison with programmed standards, and larger AMT control systems (*see* p. 42). These run manufacturing operations in, for instance, the engineering, chemical, food and pharmaceutical industries.

Optical recognition and image processing systems have proved to be very successful applications of computers. The latter is used, for example, to generate the special visual effects which are used by the television industry. The range and complexity of the effects they can produce is quite stunning. A totally different application is the visualization of soft X-ray images of baggage at airport security checks; the image can be processed to show up items of different densities in different colours.

A similar principle is used for diagnostic purposes in the processing of images from whole-body scanners used in hospitals. In a very new development of this, the image is processed and linked to a robotic system in which a robot, controlled by a surgeon, uses the detailed information from a scanner to perform neurosurgery. However bizarre this may seem, the accuracy with which the robot can operate and the degree of completeness with which it can excise a tumour (a critical factor) exceeds the skill of the surgeon.

One other area which looks ripe for exploitation in the next five years is that of domestic control. The plummeting costs of VLSI (Very Large Scale Integration of processor and memory chips) now makes it feasible to develop systems that will enable householders to undertake a whole variety of tasks by computer, ranging from controlling the central heating to paying the bills. Such systems should also be able to be operated remotely via a modem, so that changes, checks and control of security etc., can be made by the householder when he or she is away.

It is the market rather than the technology which is important in determining viability (*see* Chapter 1 p. 18). Microcomputers and home knitting machines, for instance, have both been in existence for some years. The person who had the idea of controlling the pattern of a knitting machine with a small computer was able to satisfy a significant market with a product substantially better than anything which had gone before.

■ SOFTWARE DEVELOPMENT

Most of the applications above require their own specific software, and there seems to be an undiminished opportunity for novel software of all kinds. New and faster processor chips and the increased RAM (memory) and disc capacity of current computers make new and enhanced desktop applications feasible. The market still seems to be able to absorb good quality software developments. There is a constant technical and commercial battle to bring out better ways of controlling and interfacing with a personal computer. The competition between OS/2, Unix, Windows, etc. for the best operating system and GUI (a Graphical User Interface like 'Windows') remains fierce.

At the front-end are developments in 'self-learning' programmes, artificial intelligence, expert systems and 'fuzzy logic'. Some applications of the latter already exist, especially in robotics (*see* p. 41), and it seems likely that a number of graduates and academics will find commercial success in these areas.

There have been significant developments in use of bar code readers and electronic point-of-sale transactions which integrate stock control, monitoring sales and the actual financial transaction (usually via a cash card). These, originally applied to large supermarkets, are now increasingly used by garages and smaller shops. The firms designing, supplying, installing and maintaining these systems are not all part of multinational computer companies: many started in quite a small way, having identified the future market early.

The commercialization of so-called 'smart cards', cards which contain a chip and a small battery, is now also taking off; these can contain information about a person, his bank account, medical history, codes for security clearance, etc.

There are, however, still many opportunities in the development and application of more conventional software. The most obvious examples are in general office and business systems, improved desktop publishing and accounts packages. These are very competitive areas, so your idea will have to be a good one if it is to compete successfully. There is still a need for systems to be simplified, so that memory management, multi-tasking and control of peripherals like printers do not demand increasing knowledge of technical matters.

The less universal applications may well be a better bet for a new business. You may have worked in a hospital, a transport firm or in a school, and recognised an opportunity for a computer application which does not at present exist. It could be a self-diagnosis program or a means of keeping track of the number and type of meals needed by the patients; or a system for minimising journeys for multiple collections and deliveries of the transport firm; or a teaching program for remedial maths in the classroom.

These are all real examples of programs written and successfully commercialized. They were successful because they were *market-led* innovations. The fact that the entrepreneur actually worked in the field was the key feature of detecting, understanding and satisfying the market.

■ ROBOTICS

A computer becomes a robot when it is equipped with sensors and actuators. The sensors detect information from the environment, the computer processes the information and sends signals to the actuators so that a response can be made. In the 'classical' robot this would be a movement of a robotic arm, but it can equally well respond by turning on a light, altering the temperature of a furnace or adjusting the pressure or acidity in a reactor vessel.

Note that this implies that their sensory characteristics can be much broader than a human's. They can hear ultrasound, detect infra-red or ultra-violet radiation, recognise a magnetic field and detect the presence of a specific vapour in the atmosphere.

Robots can be integrated into many kinds of system. They have found application in medicine, where implanted robots can sense neural impulses and respond by mechanical movement of some kind. Such devices can be of enormous assistance to the handicapped, since the person's own nervous system can be used to move prostheses or work a computer which can then do a whole variety of other things.

A grimmer application is the remotely operated vehicles and handling devices used to inspect cars or buildings suspected of being booby-trapped, and then disarm the device.

They are also employed to make the microchips which they themselves use. The designs for the chips are developed using a CAD-system (*see below*), and they are manufactured by a *microfabrication* process which reliably and reproducibly etches millions of electronic components onto square-centimetre slices of silicon or germanium oxide. This technique increases reliability and reduces costs.

Current developments are principally of the sensors and software. They are being taught to recognize the spoken word and sometimes the speaker, too. They are being trained to recognise three-dimensional objects, and to place and rotate them to a particular orientation. Thus, the machine is more able to *perceive* as opposed to simply receiving information.

Besides its manufacture, the modern motor car owes much to robotics. Most now have an electronic sensing system which monitors and controls fuel injection and timing of ignition, anti-lock brakes and even suspension. Modern buildings are also robotically controlled, using sensors and feedback mechanisms to govern heating, lighting and security.

The applications of robotics are thus legion. Many firms are specialising in systems design, component design and manufacture and are developing specific sensing devices to meet this need.

■ CAD/CAM AND AMT

The computer has revolutionized much engineering design and many manufacturing processes. Computer Aided Design (CAD) is now employed for everything from architecture to designing microchips and electronic circuitry, and mechanical engineering design. The software for this has evolved from simple two dimensional design programs to suites of programs allowing the generation and manipulation of three-dimensional designs and perspective views.

Linking a CAD system to a computer-controlled machine tool results in a system for Computer Aided Manufacture, or CAM. This is, in fact, a specific kind of robot: one which has been programmed to fabricate a particular article, or part of it. The program containing the dimensions of the article and the various machining stages which have to be undertaken on the initial block of material are contained in the computer code, and the machine tool responds to the code by actually performing the machining tasks. Computer Aided Manufacture/CAM has revolutionized many aspects of the manufacturing industry, and has also altered the type of training needed by people embarking on a career in it. Whether the factory is producing elaborate furniture designs or machining gears, CAD/CAM has made it possible to achieve complexity in design with reproducible quality at a low cost.

Taking this concept one stage further leads to Advanced Manufacturing Technology or AMT. This involves the complete automation and control of a whole manufacturing process, from the specification and input of the raw materials, through the process itself to the end product. All the parameters involved are continuously monitored and kept within norms. Such a concept can apply throughout manufacturing industry, to chemical processes just as much as to motor cars. The whole factory, in other words, becomes a robot.

☐ One expects to see this sort of thing in a car factory, but one of the most impressive systems I have seen is a pasta factory in Southern Italy, which was capable of automatically manufacturing a vast variety of shapes, sizes and consistencies of pasta products.

The starting point was the compilation of the order requirements from customers – so many packs of spaghetti, so many of macaroni and so on. This information was fed into the computer system which then calculated the quantities of the durum and other types and grades of wheat and other materials required to fulfil the orders. These were not constant, since the prices of various cereal strains from different sources varies from day to day and from season to season. The computer calculated the optimum mix for quality with the minimum price. This information was fed back to the financial controller, who thus had access to daily information on prices, costs, possible lost production, etc.

These quantities were automatically weighed, mixed and cooked. Different sub-assemblies incorporating modified extruders produced the required quantities of the different goods, which were then automatically weighed and packed into separate customer packs. These were collected – by robots, of course – into larger boxes, which were sealed and labelled with the address of the customer.

The boxes were conveyed by a system of robot trucks to storage accommodation, and the position of each pallet of boxes was recorded by the computer. Finally, when each customer's van arrived at the loading bay of the factory, the computer was able to 'call up' the various boxes of different pasta required to fulfil a given order from throughout the warehouse, and these were conveyed from the storage area to the loading bay by the robot trucks.

There were *some* humans around! Sometimes things do go wrong, and even automated machinery needs maintenance. There was a control gallery, where a relatively few system controllers were able to monitor every part of the whole operation on a battery of screens.

Finally, the manager of the factory had a console in his office from which he could call up any of the information available, giving him a constantly updated picture of his entire production process, from order to dispatch. □

The design and implementation of AMT is not something which can be bought 'off the shelf', and it involves aspects other than sheer technology: frequently, the whole management structure has to be rejigged and retrained. A number of major management consultants have started specialist groups to advise manufacturing firms on this subject. However, smaller firms of experts have also made successful inroads, often led by an academic specialist in robotics or automation who moves out of a university into a commercial environment.

Telecommunications

It is impossible to separate out many of the areas of current technological application into water-tight compartments; some products or systems invoke several areas of new technology. Thus, many electronic and computer systems involve or impinge on telecommunications technology.

Although satellite communication has excited the public interest more, the development of optical fibres for data transmission has also been a real success. Developments in the chemistry and manufacture of these to a rigorously high quality has been a major step forward in materials science (*see* p. 35).

Their operational success lies simply in the vast amount of information each can carry simultaneously. Techniques of removal of redundant data, data compression and bundling information into 'packets' which can be 'multiplexed' or interleaved at intervals of a few milliseconds before transmission, all continue to improve the performance of telecommunications systems.

To give an indication of the capacity of an optical fibre network, the current electronics used in the UK telephone system digitises information (i.e. converts it into electronic code) at a rate of 64,000 bits per second before multiplexing it. Thirty such signals are then merged by multiplexing to generate a signal carrying about 2 million bits per second. These can be further concentrated for transmission by optical fibres: British Telecom uses fibres carrying 565 million bits per second and has tested systems running at 2,400 million bits per second.

So important has this topic of optical transmission of data been, involving the modulation and transduction to and from optical to electronic signals, that a whole new topic of 'optoelectronics' has emerged. This is now taught as a discrete subject in many electronic and engineering degree courses.

Major companies make the fibres and install the networks, but there are still many openings for the smaller firm. These could be at the technical end, perhaps improving the amplification of optical data without the need to convert to electrical form and then reconvert, or enhancements in the method of switching. Or it could be a method of improved data compression.

Other recent innovations in telecommunications have included digital exchanges, cellular phones and the ubiquitous fax machine. We can now bank by phone, and conduct international teleconferences using the telephone network. We can work from home, using the computer and the telephone link to run a home-based office, perhaps by electronic mail.

Work is already underway on pilot schemes which will extend fibre optic communications right down to the individual home. This will enable subscribers to receive or transmit a range of video, voice and data communications channels. The economics of demand and scale will determine the commercial feasibility of such schemes, but if they do prove potentially viable, there is ample scope for the development and marketing of innovative peripherals.

Some of these innovations are being developed using the enormous resources (technical as well as financial) of major multinationals, but others have been successful products of small technology-based firms. Can you spot a market for the next development and have you the technical know-how to develop it yourself?

Electronics

Computers, AMT and telecommunications all depend on electronic principles. Conversely, many industrial control and instrumentation systems depend on logic chips and programmable logic controllers. When such control systems for manufacture are integrated throughout a manufacturing process, they become AMT systems.

A number of consultant and service companies who have grown up in the control and instrumentation area, make a living by advising on the design, installation and maintenance of systems for specific unit processes within manufacturing industry. Their customers are frequently the smaller firm which has no need to upgrade to a comprehensive AMT system.

One can extend the field into electronic gadgetry of all kinds, ranging from (to take recent examples) infant-breathing monitors to guard against 'cot death' to domestic security systems. The defence and avionics markets also continue to offer opportunities for electronic innovation.

There is an enormous market for domestic electronic products. The major multinationals are working on high quality television (HDTV), improved hi-fis and digital video recorders and tape recorders. There is much confusion in some of these areas due to the lack of an internationally-accepted standard and much money, time and effort will probably be wasted as a result of these delays.

The smaller firm will probably not be able to compete with the Sonys of this world in designing and developing a 'Walkman' or a lightweight video camera (although I should be delighted to be proved wrong!). They *can* however compete in the market for peripherals. For instance, the advent of the home video camera has led to several home-editing systems from smaller firms which are now in the shops.

Energy

It seems unlikely that many readers of this book will be contemplating a new design concept for a nuclear power station. Nevertheless, alternative forms and increased efficiency of energy generation have all been successfully developed by the individual entrepreneur. Some are noted in more detail in other sections, such as conversion of waste into fuel by chemical or biotechnological means (*see* p. 53).

■ ALTERNATIVE ENERGY

We hear a lot about 'clean energy' and 'alternative energy' these days. We are increasingly aware that the damage caused by the current scale of burning of fossil fuel (oil and coal) could affect the whole planet in the long, or even the medium term. If the politicians eventually pluck up the courage to impose a 'pollution penalty' or 'carbon tax', some alternative forms of energy production will become more financially attractive. There seems increasing hope that some of these systems *will* enable significant amounts of energy to be produced at low cost in terms of both cash and environmental damage.

It has been left to the smaller firm or the individual 'champion' to demonstrate the feasibility of wind or tidal energy in the UK. Solar energy is now being used to a small extent in British homes, but its widespread application remains principally in sunnier climes.

One must keep these developments in perspective. Electricity generation by wind or solar energy has so far produced only a relatively minuscule fraction of energy needs. Nevertheless, the changing basis of costing generation means that such small units could become a practical possibility in the future, especially in remote or outlying areas. It is worth noting that California has around 13,000 wind turbines, and Denmark over 3,000, which produces some 2.5% of its electricity: not an insignificant saving in fossil fuel.

There are other ways of harnessing natural energy to our use, besides the generation of electricity. The elegantly simple and rugged water pump recently invented, which uses the flow of the river itself to power it, is an excellent example of innovation which could improve irrigation in developing countries.

■ ENERGY EFFICIENCY

Much work is currently directed, both in academic and industrial research, towards the more efficient use of energy. One aspect is the use of better insulating materials, which may be applied both to insulating homes and factories and to cladding furnaces and reactors. In this sense, some of the developments in refractory and highly insulating materials have enabled significant progress to be made.

Another is the development of improved batteries and fuel cells. Developments in other areas of technology such as home video cameras and 'notebook' computers have fuelled the demand for longer life batteries. Current research is directed to the use of doped polymers as electrodes and also as replacements for the electrolyte solution.

Similarly, but on a larger scale, the development of the electric car, seen as a means of reducing pollution in large cities, is still basically limited by the storage capacity of the current batteries available. This goal has led a number of individual scientists to set up companies to exploit patents which they have secured on battery or fuel cell technology.

There have also been a number of attempts to develop and set up production of new internal combustion engines to supersede the conventional four-stroke engine. Many of these have been based on sound engineering and thermodynamic principles (frequently, they have been based on the Stirling Engine, which would appear to deserve a better fate than the relatively minor uses to which it has been put). Unfortunately, they require the jettisoning of all the technology and development of the four-stroke engine which has taken place over the past 80 years, a complete re-tooling and retraining of the production lines and personnel, and a total re-education of garage mechanics.

This is too much for industry to bear. It probably implies that either an innovation on this scale has to be a stepwise improvement of the current product or it has to be very much – twice, or ten times – better than the existing product. A five or ten per cent improvement is not worth all *this* trouble.

Chemistry

Chemistry is the least tangible of the sciences, yet its impact on our lives has been far greater than many more obviously technological products.

For instance, we are impressed with our new car, with its aerodynamic shape, smooth ride and computer-controlled ignition and fuel injection. We can readily appreciate the engineering and electronic skills involved.

There is probably at least as much innovation in chemistry in it as well, however. The shiny new paint has significantly better 'chip resistance', thanks to new polymer and coating technology. The smooth-running engine is maintained in good condition by lubricants specially developed for the purpose. The rubber in tyres has been improved substantially over the years. The petrol used has been refined by processes using chemical technology and catalyst development; and various additives introduced to enhance its performance have been chemical innovations. Even the wear-resistance of the upholstery involves materials developed by plastics and polymer chemistry.

One could take almost any household object and analyse the chemistry in it in this way, and this would still exclude the major developments of pharmaceutical and food chemistry. Chemistry impinges on virtually every other area of technological development, from the surface coating of microchips through the development of biodegradable plastics to the purification of genetically engineered proteins (*see* p. 52).

The point is that there is an enormous market for chemical products, and still substantial scope for development. The individual entrepreneur can still succeed in this area as in most others. Frequently, they have experience of working in a technical capacity in a large firm, spot a market or develop an idea which their employer has no wish to pursue, and decide to start on their own.

Development of water-based paints for specific purposes, surface treatment of glass to render it easier to clean, novel applications of photochemistry for pharmaceutical production and an improved formulation for a pet shampoo and insecticide – these are all examples of recent developments within smaller firms involving chemistry and new chemical products.

Pharmaceutical and medical products

The health-care market is an enormous one, ranging from over-the-counter medicines to whole-body Nuclear Magnetic Resonance machines. Both these areas have been exploited by entrepreneurs and, indeed, the establishment and growth of the UK company Oxford Instruments to exploit applications of NMR is an excellent example.

Besides the obvious areas of new pharmaceuticals and medical devices, there are many other opportunities for the entrepreneur. The development of diagnostic kits is one such area, with the exploitation of antibody technology: this is dealt with under Biotechnology on p. 53. There has also been an increase in the number of consultant companies, handling things like clinical trials on behalf of pharmaceutical firms and advising on quality assurance in drug manufacture.

■ PHARMACEUTICALS

Development of new drugs themselves is expensive, and, if a university group does discover one, the subsequent commercialization has to go elsewhere. The costs of developing it, safety testing to DHS and FDA legislative requirements and then marketing it are prohibitive to all but the major pharmaceutical companies.

Nevertheless, the academic research environment and input is vital for totally new developments of this kind, and some major multinationals have recognized this by participating in joint ventures with universities or their departments, with the company providing the cash for the research and the university assigning the rights of exploitation (or, at least, the right of first refusal) to the company.

It is difficult to predict likely winners for the next decade. Ten years ago, the scientific literature abounded with papers on the potential of liposomes and antibody-drug complexes for the targeting of pharmaceuticals; in other words, they could theoretically deliver a higher concentration of the drug to a particular site in the body, thus enhancing its therapeutic effectiveness and reducing toxic side effects.

These inventions have not lived up to their implied promise. Undoubtedly, there are applications where they have enjoyed significant success, but they have not yet measured up to their originally perceived potential. It now seems possible that monoclonal antibodies can be used more effectively for things like targeting tumours, and we may yet see the potential of this concept realized.

■ MEDICAL DEVICES

Developments in medical care do not necessarily have to come from a prestigious university or research group. In 1991, Daryl Robb, a builder's labourer in Australia, invented a retractable needle for hypodermic syringes which can reduce the risk to health workers of infection with AIDS or other transmissible diseases. Many other groups have previously attempted to develop such a device, but his design is now claimed to be the best way of preventing being accidentally pricked. It will be interesting to watch the progress of this invention, since the market for such a syringe is estimated to be around six billion units per year.

Many medical devices have been developed and successfully marketed by smaller firms. 'Devices' can be very broadly interpreted, and can include surgical implants (sometimes incorporating new materials), electronically controlled prostheses, the whole range of medical electronic equipment from anaesthetic machines to nurse call systems and the peripherals and ancillaries to accompany all these.

I am impressed by the inventiveness of medical and health care staff, who seem adept at producing practical solutions to everyday problems they and their patients have to face. Frequently, these solutions are of the 'sealing wax and string' variety, but they are usually cheap *and they* work!

In my local Health Authority in Clwyd, North Wales, I have seen purpose-designed surgical instruments to assist in particularly tricky procedures during operations, novel chair lifts for ambulances, new games to help the chronically sick child and computer systems which provide a new dimension to the life of disabled children.

There is no reason why such inventiveness should not be typical of the staff of many Health Authorities, and many of the ideas could be considered commercially. This is not to imply that all these good people should leave their chosen profession and become entrepreneurs, but it does suggest that entrepreneurs looking for new products in health care might do well by tapping into this source of invention. Many of the inventions have general applications, sometimes outside the health care field.

Biotechnology

In one sense biotechnology has been with us for a long time. Our exploitation of micro-organisms to produce bread, cheese and various forms of alcoholic drink date from antiquity. We also learnt to alter or improve the properties and characteristics of plants and animals by cross-breeding long before anything was known about the structure of the gene itself.

However, the past 40 years has witnessed a major increase in our understanding of biological processes at the *molecular level*. It is this increase in knowledge which has in turn led to the development of the tools and techniques to manipulate and control these molecular processes, in one sense treating biology as an extension of chemistry. The application of this understanding and these techniques is what we now understand by biotechnology.

It was a technology which all the pundits saw coming in the mid-1980s. It was hailed then as a revolution waiting to happen, with immense potential for commercial gain. Much money was invested as pure speculation, and the enthusiasm of the investor was fired more than a little by some researchers who were eager to promise rapid rewards for some research support.

The bubble burst in the late-1980s and many lost their investments. It was a salutary lesson and it has had the unfortunate result of partially inhibiting further investment: 'once bitten, twice shy'. Why did this occur?

There seem to have been several reasons. One was simply that the researchers promised too much, too soon. Investors simply did not understand the technology but became starry-eyed at the rich promises offered. Also, the researchers frequently underestimated the problems of scale-up. Clones (i.e. groups of similar cells) which grew well on the laboratory scale refused to grow in larger vessels, or lost their special characteristics after a few generations. Overcoming the physicochemical and control engineering problems of maintaining large cultures of cells needed some time. So too did the difficulties of work-up, that is, of separating and purifying the product from the whole mass of culture.

Those involved in developing transgenic species (such as nitrogen-fixing cereals) found unexpected scientific problems. They also found that the release of such species into the wild raised ethical questions. In some countries public opinion against such release has increased rather than diminished with time.

The preceding paragraphs have been included to illustrate the gap between the research concept and its exploitation. Some hard lessons were learnt, much money was lost and many fledgling biotechnology firms failed to take off.

This does *not* mean, however, that biotechnology has no further promise, and that it was all 'hype'. Far from it. Several new products are now in full production, and the projects which are now underway are generally much more realistic and better considered. They are also now backed by a considerable amount of scale-up experience. It seems likely that the rewards of biotechnology are still there, but they will take longer to reap than anyone expected ten years ago.

■ GENETIC ENGINEERING

Genetic engineering is a bit like using numerically controlled machine tools (*see* p. 42) but on a molecular scale. The gene may be regarded as a piece of code which specifies a finished product (a protein molecule) and the machine tools are in the cell itself (its organelles such as ribosomes with their power units, the mitochondria), which are able to take the code, read it and fabricate precursor molecules into the finished product.

The protein product of the process may be the final product the cell wants for itself or for export, or it may be an enzyme, a catalyst which is able to recognize other molecules and convert them into the ultimate final product.

These techniques have been used by biochemists to synthesize a variety of hormones and physiologically active products. As an example, it is easier in principle to isolate the gene coding for insulin, introduce it into a cell culture and harvest the product, than it is to extract porcine insulin in the traditional way from large numbers of pig or cattle pancreases. The high added value of the product and the large market makes such a process commercially feasible, but it is important that the bio-engineered product should have properties which are as least as acceptable as the natural one.

Genetic engineering can be applied to the cells in whole multicellular organisms as well as to cells in culture. Current work ranges across an immense variety of applications. These include developing and breeding infertile strains of larvae to control pest infestations, cloning varieties of laboratory animals which can form good models for human diseases and introducing nitrogen-fixing genes (from naturally occurring bacteria) into strains of cereal to potentially obviate the need for nitrogenous fertilizers.

Constant improvements in techniques of gene transduction are taking place, and new methods employing the splicing of the foreign gene into the host's genetic material before transduction are greatly boosting production of the foreign protein. This adaptation is already being employed in what is called 'pharming' – the use of cattle, goats or sheep to produce human proteins for pharmaceutical use.

Similar procedures are now undergoing their first tentative tests in humans, on patients with inborn errors of metabolism, that is, they are born without a particular gene: sickle-cell anaemia and phenylketonuria (PKU) are perhaps two of the better known ones. It may prove possible, maybe even by surprisingly simple, apparently naïve means, to introduce the missing gene into the patient's own cells. As these cells proliferate, so does the number of copies of the gene, thus eventually remedying the inborn deficiency.

To illustrate the breadth of application of genetic engineered proteins, scientists at the CSIRO Division of Animal Production in Sydney, Australia, have developed a technique known as 'biological wool harvesting', which effectively means that sheep can be made to shear themselves! They inject the sheep with a genetically engineered version of the protein epidermal growth factor (EGF).

This somehow stops the production of cells that give rise to the wool fibre, but the hair follicle (around the base of the hair) continues to extrude the fibre to the epidermis. Some 4-6 weeks after treatment a natural break in the fibre appears at the surface and the fleece can then easily be peeled off the sheep.

At the other extreme of application if not of sophistication is the development of efficient and sturdy strains of methogenic bacteria which can break down biological waste into methane, thus providing a source of fuel from 'biomass'. Such reactors can be used on farms and sewage works. They are also in use to process effluent from food processing plants into fuel; for instance, spent whey from cheesemaking plants can be converted into a highly useful form of energy, which is then used in the cheesemaking itself.

■ ENZYMES, MONOCLONAL ANTIBODIES AND DIAGNOSTICS

Without delving too deeply into it, the two features which characterize 'biological' molecules are recognition and specificity. An enzyme recognizes its substrate (the substance it reacts with) and reacts specifically with it. It does not react with molecules which are remarkably similar. Likewise, a hormone recognizes and binds to its target molecule and an antibody tenaciously adheres only to its specific antigen – so that a polio antibody binds only to a receptor molecule on the polio virus (or also, just to satisfy the experts, to a binding site on the cell membrane of a particular lymphocyte clone).

These special properties of enzymes and antibodies can be exploited. Enzymes are perhaps best known as ingredients in 'biological' washing powders, but they are also used in the clinical diagnostics, and in large quantities in the food industry.

Antibodies are also part of many modern diagnostic kits. Their molecular specificity of binding is exploited to detect minute amounts of substances in biological fluids and other media. The binding is detected, amplified (often by using an enzyme system) and perhaps quantified. The trend is to develop the assay in kit form to suit the application. Frequently, easy-to-use 'dipstick' kits are produced, as in some home pregnancy test kits.

The advent of monoclonal antibodies (these are molecularly homogeneous rather than a mixture of antibodies with differing degrees of specificity for various regions of the same antigen molecule) has allowed the development of such kits with virtually total specificity and exquisite sensitivity. Diagnostic tests for food, agricultural and veterinary application have also been produced. Indeed, the diagnostic test kit has so far proved to be a major success area of biotechnology, although it will soon doubtless be overtaken by others, such as DNA probes.

A DNA probe is another means of searching for a 'needle in a haystack' at a molecular level. This uses a length of DNA (the molecule which constitutes the genetic code) from an unknown sample and compares the *sequence* (strictly, the oligonucleotide pattern from a defined enzymic breakdown of the DNA) of the four basic units of which the DNA is comprised with the sequence from a known piece. If the sequence is the same then the unknown is the same as the sample.

Such tests are now being developed to test for the presence of cancer triggering genes ('oncogenes') and for infection with specific viruses. (A virus acts as a molecular parasite, and insinuates its DNA into the DNA of its host cell; it can then exploit its host's protein synthesis machinery to replicate itself.) Another application is the 'genetic fingerprinting' process, which is able to identify the person responsible for a crime by comparing a sample of the suspect's DNA with DNA extracted from any small sample of blood, skin, saliva or semen left at the scene of the crime.

■ BIOSENSORS

This Chapter has tried to illustrate the essentially seamless development and exploitation of technology across the man-made barriers of categorizations. It therefore seems appropriate to conclude with a, perhaps unlikely, amalgam of biotechnology and electronics.

Chemical, and hence biochemical, reactions involve movement of electrons to form or break chemical bonds between atoms. Most schoolchildren are aware of the connection between chemistry and electricity from performing electrolysis of water, in which a current is passed between two electrodes dipping into water: the water is split into its constituent elements, hydrogen and oxygen (the reverse process is the principle of the fuel cell – *see* p. 47).

The point is that chemical and biochemical reactions can cause or be made to cause detectable electrical changes which can be amplified and measured. If such a change was the result of one of the highly specific and sensitive reactions discussed in the last Section, we then have a way of developing a 'biosensor', which can monitor a particular substance and, by suitable electronic transduction and amplification, continuously display or record its concentration.

The promise of such biosensors for monitoring such things as blood glucose, drug or metabolite concentrations has led to a large amount of research and development in the past decade. It must be admitted that this is another area where the reality has not yet lived up to the promise. Many biosensor systems have worked well in the research laboratory, but the development of prototypes into reliable, rugged and durable products has proved very difficult.

Some sensors are now available which are acceptable as production instruments, and there has been steady but slow progress in overcoming the practical difficulties. There seems little doubt that they will appear within the next few years.

Chapter 3
PATENTS AND INTELLECTUAL PROPERTY RIGHTS

The concept of intellectual property

This book is not intended for the lawyer or patent agent. This Chapter endeavours to keep things as relevant and straightforward as possible, but it will soon become apparent that the subject is *not* straightforward. It is easy to become bogged down with jargon and legal terms when dealing with intellectual property rights, all of which serve only to confuse the entrepreneur. The complexity of even minor matters adds to the mystique of the specialist and, of course, to the fee charged.

On the other hand, the complexity of patent law and the skill required in drafting a patent makes the advice and expertise of a professional agent important when it comes to considering or making a patent application.

From the innovator's viewpoint, Intellectual Property implies any concept, device, design or invention which has been conceived or constructed by the innovator and over which he wishes to claim some kind of 'ownership' as his own 'property'. The 'Rights' come in when he wishes to establish some kind of prior claim to the Intellectual Property so that he can choose to do something with it. In reality, the 'Rights' do not so much give him the right to do something as *prevent* others from doing so.

So, even at this early stage in the Chapter, the plea must be made to the entrepreneur to *take Intellectual Property Rights seriously*. You would not purchase a computer or any other tangible asset for your company without protecting it by insurance. So why be slow in protecting your intangible assets, which just might prove to be immensely valuable?

The first rule is simple: *keep it secret*. This does not just mean keeping it from your obvious competitors. Even mentioning your idea to a colleague over a coffee can be considered as a disclosure, which might render any attempt to secure rights invalid. So, if you do have a good idea, make sure everyone you discuss it with understands the confidential nature of the information – preferably, they should sign a secrecy agreement (*see* p. 83).

The second rule is to seek advice; at first, maybe not from a specialist patent agent, who may well tell you to file an application immediately whether this is the best course or not. Instead, seek help from the local Enterprise Agency, Business Advice Centre, Innovation Centre or Science Park. These matters will be explored in depth later on.

The third and final rule is to seek advice early; many inventors have lost the rights to good, exploitable ideas by doing nothing about protecting them until it is too late. Do not be frightened of going to a patent agent and being told that the idea is not patentable. It is far better to do this than to leave it for months only to be told that someone else has just filed to protect the same invention.

There are various sorts of intellectual property and various ways of protecting them. They include:

- Patents
- Registered Designs
- Trade Marks
- Copyright
- Unregistered Design Rights
- Confidence.

The Law relating to them is principally enshrined in The Copyright, Designs and Patents Act, 1988 (usually referred to as the '1988 Act'). Other relevant legislation includes the Trade Marks Act, 1944; the Registered Designs Act, 1949; and the earlier Patents Act of 1977. Legally, intellectual property is a complex subject, which is why specialist advice is important.

Before we delve into the complex subjects of patents, copyright and design registration, let us first consider the more basic topic of confidential information and know-how in general.

Confidential information and know-how

Confidential information, know-how and trade secrets are assets to a business, and frequently vital ones. Although there are no specific statutes dealing with these subjects, their importance is recognized in common law.

Even when you get to the stage of telling people in your company or circle of confidants about your idea, project or design, it is essential for them – and you – to protect your interests by recognizing that the information involved has a commercial value. It is also vital that secrecy be maintained to prevent any claims of prior disclosure which might invalidate a patent.

This is particularly true in a university environment, where there is a natural and – from the academic viewpoint – laudable desire to impart, discuss and publish new ideas and knowledge. Nevertheless, the fact remains that prior disclosure or publication can render valueless all subsequent attempts to obtain a patent.

Academic institutions are recognizing the importance of maintaining secrecy on commercially sensitive research, at least until any intellectual property has been protected. Indeed, commercial organizations sponsoring such work usually insist on it.

As someone with previous research interests which have led to several patents, it may be of some help to the academic reader to learn that the progress of my research has never *suffered in any way from the relatively minor delays in publication necessitated by the acquisition of priority dates for patent applications. Indeed, the willingness to accede to these short delays has usually ultimately resulted in more financial support for the work than I would otherwise have got. In the longer term this has resulted in more publications rather than fewer.*

■ PROTECTION OF SECRET INFORMATION

The first point here is to be serious but not obsessive about secrecy. The second is to be clear about exactly what it is you want to keep secret. If there is a leakage of this information, it will be very difficult, if not impossible, to prove that it is still confidential.

Secrecy is usually a contractual matter. The simple Secrecy Agreement appended to this Chapter (*see* p. 83) can suffice. It is also far more tangible than simply telling your colleague to keep something secret, and its mere presence and his signature to it is an indication to him that you 'mean business'.

Employees and employers

Even if secrecy is not expressly stated it can be implied. It is thus the duty of an employee to keep secret and confidential information imparted to, or discovered by, him during the course of his employment. Employers should note that it is highly preferable for such matters to be overtly stated in contracts of employment of their staff.

Clients

Use of a simple Secrecy Agreement is necessary when negotiating the sale, use or licence of intellectual property. However, every copy of your product may itself contain confidential information, and in some cases, such as computer programs, every customer is asked to enter into a copyright or licence agreement with the company. The usual way this is done is to restrict the use of the product to a specified place or department, or to include some means of clearly displaying the copyright.

Computer users will be familiar with the copyright warning screens which are incorporated into most proprietary software. They will perhaps be less familiar with copyright protection sub-routines actually incorporated into the programs. These can either prevent copying, erase the program or simply be redundant lines of program which, on inspection for possible infringement, can only have occurred in the suspect copy by copying from the original.

Patents

■ THE BASIS OF PATENTABILITY

A patent is the most familiar type of Intellectual Property Right. It takes the form of a legal document by which the State gives the inventor a monopoly on his invention for a period of time – 20 years in the UK. In return for this the inventor has to disclose exactly what his invention does and how it works.

Patenting is expensive, especially if patents are required in many countries, as they frequently are. They can also take some time to be granted.

Patents are granted on the following conditions, *all* of which should apply.

A process or product

You cannot patent an idea in general, but only an article, physical 'invention' or process, which is capable of industrial application. Computer programs themselves are not patentable but products and processes which are computer related can be patented. However, programs *can* be protected by copyright (*see* p. 73).

Novelty

The process or product must be novel at the moment when application is made, and not simply a minor modification of one already in existence (if it is, it can be deemed 'obvious' and thus not patentable

– *see* below). If you believe that others may be following in your footsteps, then undertake a literature search to confirm the novelty of your invention, and see a patent agent with the intention of filing a Provisional Application (*see* p. 67) immediately.

Unobviousness

This relates to the previous point, and again prevents patenting of a trivial modification of something which is known. There must be an identifiable 'inventive step'.

The definition of obviousness is subjective and its interpretation is often left to the patent examiner. However, one test of obviousness is whether anyone else 'skilled in the art' (i.e. someone who is a professional in the same technical area) would reasonably describe the invention as trivial.

For example, if you were contending that your new three-paddled stirrer blade was a significant and non-obvious development over a two-paddled one, you would fail. If, on the other hand, you explained that you had designed the angle and shape of the blades to optimize stirring efficiency and minimize cavitation, and produced mathematical and design information to demonstrate the significant advance inherent in your system over all others, it might well be patentable.

Industrial applicability

The invention should clearly have industrial utility, and should not involve application of impossible scientific or technological principles. The term 'industry' is applied in its broadest sense of any practical activity.

The purpose of this is to ensure that any patentable invention has physical characteristics of a technical nature, and belong to the 'useful or practical arts', as distinct from the aesthetic or fine arts.

Exclusions

The essential feature for patentability is thus an *invention* rather than a *discovery*. A new mathematical theory, an artistic or literary creation or a new format for presentation of information are thus not patentable (although some can be protected by copyright).

There are certain other exclusions, such as a method of diagnosis practised on the human or animal, or a means of treatment by surgery.

■ WHAT DOES A PATENT DO?

A patent prevents *others* from doing something: it confers no additional rights allowing the inventor to do something which he did not have in the first place. It gives you, the inventor, a monopoly on your invention, and the right to take action if others try to infringe this right. You therefore have the sole right to the manufacture, use or sale of the invention within the country for which the patent is granted. In the UK, the period of validity is 20 years. To maintain it in force, there is an annual fee which has to be paid from four years after the date on which the patent is filed.

Whether or not you actually decide to make and sell the invention yourself is up to you. It may be preferable to sell or 'assign' the patent outright to another company. Or it may be better to permit another company to exploit it for a fee; this is known as *licensing*, and can be an excellent means of generating income from your invention if you do not wish to make and sell it yourself.

A patent provides a clear, legally-acceptable description of the invention, and it is thus apparent to others exactly what is owned. Since a patent is tangible evidence of originality, it can indicate to others that you have an intellectual asset which could be exploited. This can, for instance, influence a bank or investor to lend you additional finance. Equally, you now have something to sell to others to exploit. Despite all this, it is usually easier to obtain a patent than to successfully exploit one.

Although the monopoly bestowed on an inventor by granting of a patent is enforceable by law, it is the patentee's responsibility to maintain his rights. The Patent Office itself will not detect infringements and take legal action. This is really the 'down side' of patents, and it can be a problem for the small inventor, with only limited time and finance.

The competitor might, for instance, be a big company with massive legal and financial resources. It may make a counter claim against you, stating that your patent is not valid. Or it may be claimed that your invention was not novel, but obvious; or that it has already been published or used. However sound in law your case may appear, the time and legal costs in defending it can be prohibitive.

■ PATENTS AND NEW TECHNOLOGY

The age of technology has stretched the theory and practice of patents to the limit. Paradoxically, only the advent of information technology, optical disc and other fast data retrieval systems has enabled patent offices throughout the world to even attempt to keep up with the immense number of applications being filed, considered and granted.

Incidentally, there is no such thing as a 'World Patent'. The best we can currently do is to operate under the Patent Cooperation Treaty, which allows a single patent application to cover 48 countries. Before the Treaty, signed in 1978, applications had to be filed separately in every country in which protection was sought, which was immensely expensive and cumbersome. Do not, however, imagine that having the Treaty makes the matter simple.

New technological developments have also stretched the legal basis and definitions of patents to their limits, particularly in software and transgenic organisms. If you have a potential invention in these areas it is important to get good expert advice, since the arguments, especially in biotechnology, sometimes seem to border on the theological.

As an example of the current complexity, software has some of the properties of a written work, for which copyright protection is perhaps more appropriate (*see* p. 73). On the other hand, when operating the software in a computer or (say) a numerically controlled machine tool it has the characteristics of a mechanism or machine, which falls under the ambit of patent law. These hybrid features have led to substantial problems when trying to accommodate software developments under existing patent and copyright law. However, in the software area there now seems to be enough precedent for an experienced patent agent to give reliable advice.

Some attempt has been made to rationalize the problem of patenting software. In December 1990 the European Community reached a common position on a directive which should ensure the protection of software by copyright throughout the Community. This is similar to the stance taken in the US and Japan, who have also taken the copyright approach. It is hoped that Member States will now formally adopt the directive and implement it from January 1993.

The question of the patentability of life forms, however, is even more vexed. The classic case is the genetically modified mouse developed by Harvard University as a 'tailor-made' animal for scientific research. Differing attitudes, reflected in each country's laws, have been taken in the US and Europe over the patentability of such animal forms.

The fundamental legal argument is whether a genetically engineered plant, animal or tissue from them is an invention (which is patentable) or a discovery (which is not). At the time of writing, the appeals board of the European Patent Office has reconsidered its view of the 'Harvard mouse', saying it was not an animal variety, but an invention employing a microbiological process, which is therefore patentable. They made it clear that this would *not* apply to all transgenic organisms. There is still much contention over this issue, and the European Parliament have raised several criticisms.

Attempts are also being made to 'patent the human brain', as the papers sensationally put it. The US National Institute for Neurological Disorders and Stroke has a programme to sequence the DNA (i.e. the genetic code) for as many of the estimated 30,000 genes coding for proteins in the human brain as they can. What is more, they are filing patent applications for them; it is expected that they will attempt to patent well over 1,000 sequences. There is currently substantial doubt over whether any patents will be granted, since no applications of their discoveries have been proposed. The inventive step is thus not apparent (*see* p. 60).

Complications are also arising from endeavours to patent the new 'high temperature' superconductor materials (*see* Chapter 2 p. 36), with the US Patent Office having to decide which of no less than four competing applications has the prior claim. The patent lawyers are likely to get rich well before any of the inventors do.

All this makes for interesting discussion over the coffee and brandy, but the trouble is that decisions on such matters are affecting many academic and industrial research organizations, and millions of pounds, dollars and ECU are at stake. Until such time as clear decisions and guidelines emerge it is difficult to give advice.

■ WHO IS THE INVENTOR?

It is now time to take the inventor through the typical sequence involved in applying for and securing a patent. This Section considers the stages before filing an application, and the next, the filing and granting process itself.

The inventor can be a person or the company which employs him, i.e. a corporate body. The questions arise in this latter instance: who is the inventor, and who claims the rights to the invention?

If the invention arises as part of the normal work of the inventor, then the company normally claims ownership, and the contract of employment of the inventor should cover this. If, however, the invention stems from something outside the normal work of the inventor, then the individual can claim ownership. For instance, if a person employed as a tissue culture specialist happens to be a radio 'ham' in his spare time and he invents something related to his hobby, then *he* can claim ownership.

In law, for a company to own an invention, it is not sufficient that the invention has arisen as part of the inventor's normal work and duties. There is an additional requirement that the invention *might have been expected to result from these duties.* This is embodied in Statute Law and cannot be altered by any Contract of Employment.

In the past, the academic world has been fairly relaxed on ownership of invention, but increased business awareness and tight finances have led to a more rigid attitude. Most universities now include at least partial ownership clauses in contracts of employment or in policy statements on ownership of and benefits from intellectual property. However, in all cases the individual's rights to share in the rewards is maintained. If you work in an academic environment, it is as well to be clear from the outset what your authority's view is on the subject.

If you are a research student and you believe you have developed a possible invention as part of your research programme, the invention, it will generally be argued, should be jointly owned by you and your supervisor, since you would not have embarked on the area of research or progressed within it without the advice and guidance of your supervisor.

Anyone subcontracted to do a piece of research which may give rise to an invention should have the position of ownership clarified in the contract. Frequently, it is not, and difficulties can then arise.

A good patent agent will take great pains to enquire carefully so as to correctly name the actual inventors.

■ REALIZATION OF THE INVENTION

At some stage, it dawns on a person that he perhaps has an invention. It may be that he was deliberately working towards it, and that a patentable product was always a key part of the plan. Others will simply realize at some stage that the idea they have 'may be worth something', and seek to protect it. Frequently, people with good ideas need to be encouraged to think in terms of patent protection: I have had to persuade many that their idea really was worth taking seriously.

Quite often it is not advisable to apply for a patent straight away, but in other cases it is critical to do so immediately. Although a patent does give protection, it is expensive, so the ultimate rewards have to be both realistically achievable and substantial for it to be worthwhile.

The 'deal' you, the inventor, make with the State which grants you the patent also has another side to it. In return for the patent, you are obliged to disclose in full and in detail what the invention is and how it works. The protection afforded applies to the disclosures, and a competent patent agent will ensure the widest possible cover. It is thus the careful drafting of the patent which needs most of the professional skill of the agent.

Despite this, the very act of disclosure can provide competitors with an opportunity to scrutinize the application and search for loopholes. Alternatively, they may be able to see your line of thinking before you get a chance to see theirs, and they thus might develop a product to compete directly with yours.

Added to all this is the fact that it can be difficult, time-consuming and very expensive to defend your patent or to legally challenge an apparent infringement. If the infringement takes place in another country it can be even more fraught with difficulties.

So, rushing to file a patent application on every invention at the first possible moment is not necessarily the best course of action, especially for the individual inventor. As described below (*see* p. 68) there is an 18-month period between filing an application and its publication where amendments can be made.

Many inventors are advised to file to get a *priority date,* and then not only use the 18-month period to do work which substantiates and supports the claim, but also to move towards exploitation, which frequently means licensing out the invention. Eighteen months is in reality a very short time to do this. Delaying filing means delayed publication which prevents anyone else seeing the idea and then finding a way to circumvent the invention inherent in it.

If, therefore, your invention is a well kept secret it can often be better to develop the practical side of the invention, and *then* file.

On the other hand, a good idea in an area of intensive commercial activity is best filed straight away. There is always the risk that delaying filing gives a competitor a chance to achieve an earlier priority date.

■ ASSESSMENT OF PATENTABILITY

The inventor may contact a patent agent directly, or he may first seek advice from his local business advice or innovation centre. Either way, the patent agent's experience is usually eventually necessary, first to assess whether there really is anything potentially patentable in the invention, and then to define precisely what the claims are.

The patent agent, of course, has the experience necessary to avoid the various pitfalls and loopholes involved in preparing, filing and pursuing an application. He will be able to conduct searches to see if anything else has been published which would hinder the application. However, if you are a researcher with recourse to a good technical library and some experience at literature searches, you can save yourself a lot of time and money.

But beware! The patent agent's real art lies in the precise and unambiguous description of an invention, so as to identify the key areas of novelty and application. It is vital to have such a legally valid description, because its wording will be aimed at preventing others from circumventing the patent, if it is granted. If you are a scientist with some experience in writing scientific papers, it is easy to imagine that you can write such a description yourself. This is usually not so. Your own description will be a good starting point which the patent agent will find valuable and time-saving, but it will not usually be sufficient for patenting purposes.

It is not difficult to give a description of your own invention as you have made it, but if, for instance, you state the materials from which it is made, the claim would not cover the same device made from different materials. Nevertheless, the patent application must 'show reasonably' the idea it covers and must give sufficient information for the idea to be evaluated. The skilled patent agent will know how much to put in, how much to omit and how general to make the wording to afford the maximum protection to his client.

□ It is sometimes surprising what can be patented – and also what cannot. As a decidedly 'low-tech' example, I was approached once by a pharmacy technician in the local District General Hospital who frequently had to dispense medicine to blind or partially-sighted patients, and who was concerned that these patients could not read the labels on the medicine bottles. She had the idea of labelling with a stiff, plastic adhesive tape, which could be stamped to give raised 'bumps' or 'lines': one bump meant 'one tablet', one line meant 'once a day', and so on. How best to exploit her invention to benefit the patients, she asked.

Undoubtedly, this was a good idea and worthy of consideration, but, I felt, was almost certainly not patentable. Nevertheless, I checked with a patent agent, which is invariably the right course of action in such circumstances. Surprisingly (to me, at least) the invention was patentable in the context for which is was intended. A Patent Application was very carefully drafted and it has now been granted. Full feasibility trials are now under way. The moral is to always make sure! □

The process of patenting

■ THE STEPS INVOLVED IN UK PATENT APPLICATION

The process of granting a patent is a long process, taking two years or more to complete. It is by no means a simple matter of filling up forms, but is similar in complexity to negotiating a contract.

If you employ a Patent Agent, he or she will guide you through the various stages, and will remind you of the information necessary, of the forms to be completed and of the various fees involved. For a UK Patent, the cost from application to granting will probably be within £2,500 to £3,000 including the agent's fees.

In summary, the steps involved are:

- **filing** a full disclosure of the invention with The Patent Office;
- a **search** which is carried out by The Patent Office to check that the invention is new and non-obvious;
- **examination** in depth by The Patent Office to check that it meets legal and formal requirements, and is technically sound;
- **grant** of the patent, provided that no objections are raised.

The UK Patent Office has a helpful staff and publishes a number of free booklets on the preparation of and application for a UK patent. They also have an extensive Search and Advisory Service, with an on-line Patent Search and an on-line Structure Search for chemical compounds.

For a UK patent, it is to the The Patent Office that you should apply. If you need protection in several European countries, then you can apply to the European Patent Office in Münich, *having first obtained the UK Patent Office's approval to do so*. However, the usual course is to make the first application to the UK Patent Office and then subsequently apply to the European Office (*see* p. 70).

Filing an application

It is possible for an inventor to apply directly to The Patent Office for a patent, without the intermediary of a patent agent. I have seen this done a few times, and significant problems have occurred on most occasions. In one instance, it proved quite impossible to convert the patent application into a sensible, granted patent. The commercial value of a patent does not just relate to the technical accuracy of the description of the invention, but to the patent's *claims*. The wording of these claims is the vital matter, and a good patent agent will construct these to achieve an *economic* result and not solely to describe the invention. A layman can seldom achieve this.

Once you and your patent agent have defined and described the invention, the next stage is to fill in the application form supplied by The Patent Office, and send it and the description back to them with the appropriate fee. A receipt is sent back to you after filing the application, giving what is termed the *priority* date. This does not stop anyone else from subsequently claiming a similar invention (the application simply books a place in the queue), but it does make it much more difficult for them.

The effect of filing an application is to give you 12 months to decide what to do next. It is important not to waste these 12 months: they can pass very quickly. In this time, you should be busy exploring the commercial possibilities of your invention, raising finance and developing the invention further.

It is often recommended by experienced patent agents that after the 12 months a fresh application be made, accompanied by a Final Specification, and that the second application should claim priority over the first. This has the advantage of continuing the pursuance of the first application, but enables additions to be made to the description of any improvements and modifications discovered since the first filing.

Preliminary examination and search

This must be requested within a year of the initial application. On request plus the appropriate fee, the application is passed to a technically qualified examiner, who will make a limited literature search, with emphasis on previous UK patents, to check for novelty and obviousness. This usually takes three to six months. Any publications considered to be relevant will be included in a Preliminary Search Report which is sent back to the applicant.

At this stage, if there are any publications cited in the Preliminary Search Report which might cause the Patent Application to fail, you have a choice. Either you can accept that there really is prior publication, and that your invention is not novel or obvious; or that the cited publications do not invalidate your application.

You can amend the application if you wish, and therefore clarify or substantiate the original claims made. However, it is not permissible to widen the scope of the invention.

Publication

The application is published within eighteen months of filing in the Patent Office Journal. Publication is part of the 'deal' with the State for granting the monopoly inherent in the patent, and it is at this stage that others can see what your invention is and what you claim for it. They can also object if they wish, in the sense of filing observations.

It should be recognised that at this stage the patent has not been granted, and nor does publication imply that it will be. Nevertheless, it can act as a deterrent to a potential infringer.

What also happens in reality is that the avid readers of the Patent Office Journal include many who make their living by doing business with patentees. You will receive letters from people offering to take all the worry of exploiting your invention off your hands – for a fee. You will be offered insured protection against infringement, which will provide a proportion of any legal costs involved – for a premium.

The advice of a good patent agent or lawyer is invaluable in such circumstances. In some cases, it may be advisable to insure against infringement. In other instances, where, for example, the patent is to be totally assigned to a third party, it may be irrelevant.

Substantive examination

Within six months of publication, a 'substantive examination' must be requested by completing a further form and sending it with the appropriate fee to The Patent Office.

A technically competent Patent Examiner will then send a Search Report to the Applicant. Once again, there is an opportunity to amend the application, with the proviso that no attempt is made to widen the scope of the invention from that of the original application. At this point the specification is published in its full and final form.

Any dispute with The Patent Office can be queried in writing or by a hearing at The Patent Office itself. You may represent yourself or invite someone else to represent you. This process may be likened to a negotiation between the patent agent and The Patent Office to secure the best possible rights for the inventor.

Grant of Patent

If the Comptroller of Patents formally accepts the application, the patent is granted. To maintain it in force, there are annual fees to be paid from after the fourth year from application.

It should be noted that the grant of a patent does not guarantee its validity. Any person accused of infringement can apply to the

Court or The Patent Office to cancel or 'revoke' a patent, citing appropriate reasons which are laid down in law.

■ FILING ABROAD

A UK patent itself gives no protection outside the UK and the Isle of Man. It is important to weigh up the pros and cons of filing applications in countries outside the UK.

There is, of course, the expense. On the other hand, if there is a substantial worldwide market for the invention, there is nothing to stop others from exploiting it. More pertinently, for some technology-based products, the UK market is not large enough to ensure viability. In such cases, it is well worth considering the protection of additional patent cover.

Under international conventions, it is possible to claim the same priority date for applications in other countries as your UK patent, provided that:

- the application is based on the UK application;
- application is made within 12 months of filing in the UK.

This allows the foreign application to be backdated by a maximum of 12 months.

Foreign patents can be acquired by making a separate application to each individual country. Sometimes, separate applications are necessary, but certain Treaties (*see below* and p. 71) enable this procedure to be streamlined. Separate applications are, of course, expensive. Since each country has differing regulations, it is necessary to appoint an agent in each. Translations of the patent must be provided, which means employing a skilled translator who will not use terms which make nonsense of the claims or otherwise invalidate the application.

The European Patent Office

There is not yet a single Community Patent covering all European Community countries, and work is still in progress to try to achieve this objective. Although the Community Patent Convention was initiated as long ago as in 1978, not every country of the Community has yet signed it, and, indeed, there is no certainty that they will all have signed it by 1993.

However, the main European Patent Office is located in Münich. The Hague houses an International Searching Authority for the International Patent Cooperation Treaty (*see* p. 71). The so-called European patent requires an application in just one language

(English, French or German) and currently provides protection in up to 13 participating nations, each of which have to be designated by the applicant. Additional countries are considering joining the scheme. The cost of a European patent is around three times that of a UK patent.

The method of application and the sequence of procedures is similar to that for a UK patent, and, indeed, application may be made through the UK Patent Office in Newport. In recent years there have been an increasing number of applications to the European Patent Office, and consequently fewer to the British Patent Office, which has now moved out of London to South Wales (although there are still filing facilities in London).

The Patent Cooperation Treaty

As noted on page 62, there is no such thing as a 'World Patent'. There is, nevertheless, the Patent Cooperation Treaty, which allows a single patent application to cover 48 countries.

Before the Treaty, signed in 1978, applications had to be filed separately in every country in which protection was sought, which was immensely expensive and cumbersome. The Treaty is administered from Geneva by the World Intellectual Property Organization, and operates not by granting patents itself, but by conducting literature searches to assess the validity of the claim.

For UK applicants, initial application is made to The Patent Office, specifying for which countries patent cover is being sought. The international searching is undertaken by the International Searching Authority at The Hague and the Search Report goes to Geneva, to the World Intellectual Property Organization.

Increase in international trade has led to a boom in the number of patents filed with the World Intellectual Property Organization: in 1990, there were over 19,000 filings made under the Patent Cooperation Treaty, which were equivalent to over 400,000 national applications.

Defending a patent

This book is principally a guide to starting a technology-based business, and it is not intended to delve too deeply into legal matters. Nevertheless, many technology-based companies start up specifically to exploit a patent invented by the firm or assigned to it. News of infringement or challenges to the validity of the 'key' patent can come as a shock, and it is worth touching on the problems of infringement.

Sometimes the infringement is genuinely unintentional, and the mere informing of the infringer that a patent is in existence of which he is apparently unaware will be sufficient. However, this is not always so. He may have already committed himself to the expense of producing and marketing a product similar to yours. He may therefore challenge your patent, or risk the consequences of legal action, which he also knows you will find expensive.

If the infringer happens to be a major multinational firm, then in reality *force majeure* frequently conquers. Unless you have a cast iron case, it is usually best to settle out of court. One weapon which can be used effectively in such cases is the threat of adverse publicity on the larger company.

Until late 1990, all actions for infringement were heard in the Patent Court of the High Court. However, there is now the Patents County Court, also located in London, the objective of which is to develop a much more cost effective means of settling patent and design disputes.

In the Patents County Court, proceedings are initiated by a summons issued by court officials. Any defence entered must be a complete statement of the plaintiff's case, discussing and identifying all the issues. The defendant and the plaintiff are each then allowed to respond.

The papers are subsequently considered by the Judge and discussed with the parties at a pre-trial hearing. This is primarily to encourage out-of-court settlement. The case, if not settled, then moves forward to trial, which is usually short (two days or so) and is not attended by a large retinue of advisers.

The problems of defending a patent in the High Court are the marshalling of expert legal opinion and the consequent cost, which can be enormous. Cases are usually lengthy and can take up to two years to get to trial. Perhaps I should reiterate the desirability of using a competent patent agent in the first place. If he has drawn up the description and claims for the invention properly, it will be that much more difficult to infringe it with impunity – and it will also cost the infringer a lot of money to defend his action.

If you do have to fight an infringement and then win the case, the court can order a variety of remedies, viz:

- an injunction to prohibit further infringement;
- damages, based on the sum which you would have received as royalties from a licence on your Patent;
- the destruction of the offending articles;

- a Court declaration that your patent has been held valid, after which it will be virtually impossible to attack it.

The Court may also award costs, which, for the Patent Court in the High Court can amount to anything between £150,000 and £1,000,000.

These high fees, the long time before the case even gets to court, the complexity of the arguments and the uncertainty of the outcome mean that around nine out of 10 cases are settled out of court.

Copyright

Copyright law in the UK has undergone substantial reform under the '1988 Act', which covers all copyright works generated after its coming into force in August 1989.

Copyright gives a valuable and automatic protection which applies principally to literary and artistic works, and prevents others from copying them. It includes computer programs. The protection given is not for the work or idea itself but for its expression, so in this respect it differs from the monopoly protection inherent in a patent. Again, unlike a patent, the work does not have to be unique to be capable of being protected by copyright.

■ THE BASIS OF COPYRIGHT

Copyright protection is normally for the life of the author plus 50 years (*see* p. 74). The protection entitles the owner of the copyright to the right to prevent others from copying the work without his agreement.

Literary works

Naturally, this principally refers to books, and includes not only original literary works with genuine artistic merit but also compilations of information such as directories and scientific data. Computer programs are protected as literary works. Names and titles are not protected by copyright.

Artistic works

As with literary works, the question of artistic merit does not come into it. Copyright embraces original artistic works such as paintings, photographs and three dimensional art, architectural designs, and things such as diagrams, maps, plans and charts.

Technical drawings and plans from which artefacts are made can also be protected by copyright, but the artefacts themselves made by following the plans are *not* so protected.

WHAT DOES COPYRIGHT PROTECT AGAINST?

The term 'copying' is interpreted in the Act in a broad sense, which is generally favourable to the owner of the copyright. It includes:

- publishing, reproducing or transmitting the work;
- performing the work or exhibiting it in public;
- issuing or distributing copies of it;
- adapting or translating it;
- converting a computer program from one language to another;
- dealing in or importation of infringing items (provided the person concerned was aware that he was infringing copyright).

The protection afforded by copyright is normally for the life of the author plus 50 years. Note the distinction here from patents, whose duration is for a fixed period of 20 years. This emphasizes the essential feature of copyright in protecting the *author* of the work. The copyright may belong or be assigned to the author's company or another third party, but the duration rule still relates to the author.

An exception to the above rule is made for computer generated works, whose copyright lasts for 50 years from the end of the year in which it was made.

Besides the specific protections above, the '1988 Act' introduced two new *moral rights* which the author can claim. Firstly, there is *paternity*, which is the right to be named as the author of the work, provided that the author demands this right in writing. Secondly, there is the right of *integrity*, which protects the reputation of the author against disparaging or derogatory treatments of his work. This latter right does not apply to computer programs.

Although ownership of copyright can be assigned (*see below*), the two moral rights are specifically assigned to the author only.

WHO OWNS THE COPYRIGHT

The author is usually the owner, and the duration of the copyright is related to the life span of the author. If the work arises from the normal work of the author, then his company can claim ownership. Formal assignment is not strictly necessary, but it is desirable that contracts of employment should cover this.

If the work arises from something outside the normal duties of the author, then the individual can claim ownership. The author of a computer generated work is deemed to be the person who made the arrangements necessary for its creation.

If you run a company, it is important that your employees have statements in their contracts of employment that the company has an interest in protecting any invention or design which arises as a result of the work they undertake, and that it would take steps to secure such protection for the company. This can avoid friction, disagreements and possible litigation.

Statements assigning copyright should also appear in contracts for work undertaken by subcontractors, so that if you commission, for instance, a computer program, a photograph or a design, the copyright belongs to you.

Universities and colleges have generally not challenged the academics' right to own the copyright to their work. This has, in the past, generally applied to publications and artistic works with aesthetic value. However, the same may not always apply nowadays. For instance, when expensive university resources and academic time have been used to generate computer programs with substantial commercial value, the university may well insist on ownership of the copyright, or at least taking a substantial proportion of any ensuing sales or royalties.

If you are an employee of an academic institution, it is as well to ascertain the position your authority takes. The position of a professor in this respect may well be contractually different from that of a research student, a research assistant or a technician.

■ CLAIMING COPYRIGHT

There is no complex procedure or assessment required in acquiring copyright for your work. It arises automatically on production of the work. However, in practice it is highly desirable for the copyright nature of the work to be clear from the outset to any reader or observer.

You should thus state on the work your authorship, the conventional copyright symbol © and the date. Ideally, get an independent third party to countersign the work and verify the date of its existence. Such precautions are not mandatory in the UK, but they are in some other countries. In the United States, the copyright itself arises automatically, but in order to claim damages it is important to have the copyright symbol and in order to *enforce* the copyright it is necessary that the rights have been registered. In order to further complicate a complex situation, registration can be done even after a problem has arisen.

If you wish to claim copyright on a computer program or a database, then put a few 'nonsense' lines in the program or a few dummy addresses in the database. If someone then copies it but claims originality, it will be easy to show copying.

The international copyright situation is not dissimilar to that of patents: there are two international conventions in force, the Berne Convention and the Universal Copyright Convention, which enable copyright to be enforceable in most countries. However, the matter can be complicated, and it is best to seek expert advice from a patent or copyright expert.

DEFENDING COPYRIGHT

The most important matter is to be able to demonstrate your ownership and then to *prove that the infringer used your work to generate his*. This latter point is vital. It is not enough to demonstrate that your work is similar to another's. After all, two people can have the same idea simultaneously (this seems particularly true of computer programs).

The essential points are to establish:

- that it is a work in which copyright exists;
- that the plaintiff is indeed the owner of that copyright;
- that there has been an infringement by copying that work;
- that there is reason to believe that the infringer knew he had infringed.

If you claim an infringement and win the case, the court can order a variety of remedies, viz:

- an injunction on the infringer to prevent further copying;
- damages, which can be severe;
- that any remaining infringing copies be 'delivered up' to the court;
- ordering the infringer to 'deliver up' any plates, masters, moulds, discs or tapes used to manufacture the infringing copies;
- costs.

In some instances, it is not just the generation of copies which causes disadvantage to the owner of a copyright: it is the widespread *dealing* in illegal copies which causes the financial damage. If the dealer has knowledge that he is dealing in unauthorized copies, then the offence is a *criminal* one.

The computer industry is well served by the '1988 Act'. Computer programs are clearly protected by copyright, and the infringing acts include storing in any medium by electronic means and making copies which are transient. Furthermore, the renting of copies of computer programs to the public is a restricted act.

The remedies allowed by law in the protection of copyright are wide. In some circumstances they include seizure of infringing materials by the copyright owner himself. Suing for civil damages can take years and prove very costly without much benefit. Since there is a criminal liability for infringement, it is often much quicker to threaten or start criminal proceedings. There is usually a desire to avoid the stigma of criminal prosecution, and out-of-court settlements are more likely.

These examples illustrate the importance attached to copyright protection in law, and the strengths of the sanctions and remedies for infringement. The complexity of the subject and the powers available make it imperative for expert legal advice to be sought over any such proposed action.

Registered designs and design rights

A Registered Design confers protection on the outward appearance, shape or embellishment of a product which makes it more attractive. As the title implies, to receive protection, the design must first be registered at The Patent Office.

Unregistered design right is a new form of protection for original designs. Before the '1988 Act' these articles were covered under copyright protection.

■ REGISTERED DESIGNS

As with patents, the disclosure of a design to a third party except under conditions of a formal secrecy agreement could make registration of the design impossible. It is therefore *absolutely vital* that you do not talk to anyone about it until a priority date is received.

In contrast to patents, however, the 'registrability' of a design is not solely about function. Indeed, the '1988 Act' endorsed the exclusion of certain articles of a functional nature from the scope of registration. The features which make the design registrable are shape and ornamentation, and their purpose is to enhance the attractiveness of the object. The protection is to prevent competitors from making products with a *design* (rather than function) similar to yours.

For a design to be registrable it must be new, but local UK novelty is all that is required. 'Design' means 'features of shape, configuration, pattern or ornament applied to an article by any industrial process', and these features in the finished article must 'appeal to and be judged by eye'.

The owner of a registered design is normally the designer or creator, but if the design was created by an employee as part of the normal course and duties of his business, the design will belong to the employer. Similarly, if the design was commissioned by a person or a corporate body, the person or company is considered to be the owner and may apply for registration of the design.

To achieve registration, the design must be submitted to the Designs Registry at The Patent Office. Patent agents are familiar with the process of design registration, and can give advice needed. The initial cost of registration is between £250-£300, and the process takes about a year to complete. Registration lasts for 25 years from the date of application provided that the five-yearly renewal fees are paid.

Overseas registration of designs is also possible, and applications to do so must be made within six months of the date of the UK application. Many countries have a similar concept of the registered design to that in the UK, but some vary and include means of protecting the *function* of a design as well as its shape and appearance. Your patent agent will advise.

Infringement procedures, remedies and penalties are similar to those for patent infringements. One additional infringement is the production of or dealing in kits or sub-assemblies of parts which could be assembled to form an article subject to a registered design.

■ UNREGISTERED DESIGN RIGHT

Like copyright, unregistered design right is an automatic protection which comes into force upon the design of an article, provided that the design is neither trivial nor commonplace. The '1988 Act' abolished the copyright protection for articles produced from technical drawings or blueprints: copyright still remains for the drawings themselves, but not for articles made by their use.

The unregistered design right was a significant innovation of the '1988 Act'. The right exists for 10 years from the end of the first year in which articles made from the design were first sold or hired. In order to allow a reasonable time for commercialization of the design, there is an overall maximum of 15 years from the end of the first year in which the design was first recorded as a design document or an article was first made to the design.

The owner of a design has the exclusive right to use it for commercial purposes, either by the manufacture of articles to the design or making further drawings for the purpose of making articles. The design right can also be used to prevent copying of a design for use in a *different* article.

What is protected?

The unregistered design right confers automatic protection on original three dimensional industrial designs; two dimensional designs are excluded. The term *design* means the shape or configuration of any article or part of an article. Designs of semiconductor chips have design right protection, and a European Community Directive enables the rights to be exclusively available for the full 10 years; licences to copy will *not* automatically be available during the last five years of the term (*see* p. 80).

The design right will not protect:

- designs which are not original, the term 'original' implying that the design was not commonplace at the time of its creation;
- a method or principle of construction or manufacture;
- surface decoration or ornamentation;
- features of shape or configuration of an article which enables it to fit or integrate with another article so that either may perform its function;
- features of shape or configuration of an article which depend on the appearance of another article of which it is intended that the design should form an integral part.

It should be apparent that the design right is both new and complex. It is therefore highly advisable to seek advice from a patent agent or a solicitor who is competent in the field. In general, it is not wise to rely on design right if other means of protection are available.

Ownership

The designer or creator of the design is the first owner. As with a registered design (*see* p. 78), if the design was created by an employee as part of the normal duties of his business, the design is considered to belong to the employer. Similarly, if the design was commissioned, the person or company who commissioned it is *automatically* considered to be the owner of the right.

Unregistered design rights are an asset in the same way as copyright, and may thus be the subject of sale or licence agreement. One special limitation of the design right is that, in the final five years of the term of validity of the right, anyone can *by right* have a licence to commercially exploit a protected design.

Claiming the right

Like copyright, the right comes into existence automatically with the design document or the article itself. It is, however, highly desirable that a proper record or register of design drawings be maintained, and that the dates are recorded on which articles were first produced and marketed from the designs.

The design right is only effective in the UK. There are schemes available which give protection in other countries, but these are not automatic, and must be applied for.

Defending the design right

The remedies for infringement are similar to those for copyright (*see* p. 76), but many matters are settled less expensively out of court. A distinction from copyright is that, if infringement occurs during the final five years of the term of validity and the infringer agrees to take a licence, then no injunction is available, and no order to 'deliver up' will be made. In such circumstances, the damages recovered are also limited to double that which would have been payable if the licence had been granted before the infringement occurred.

It is worth noting that not everyone is entitled to use the right. The author or owner must be a UK national or a resident of the European Community. Thus, imported goods may not have design right.

Trademarks

Trademarks go on for ever: there is no statutory time limit like those for the other forms of protection described in this Chapter. The well known red triangle of 'Bass' was the first registered mark in the UK, in 1876, and is still in use today.

Under UK law it is possible to acquire certain rights in a Trademark solely by virtue of its use, without registration. However, the action (commonly termed 'passing off') to restrain use of the same or a deceptively similar Trademark is different and more difficult if the Mark is not registered. Your trade or service mark is a valuable asset, and is part of the goodwill of your company. If you have a trademark, it is well worth registering it.

■ THE BASIS OF TRADEMARKS

A trademark can be a word or symbol (the old-fashioned term for 'logo') which *identifies* the company or its services and distinguishes it from other traders. Use of a registered mark gives the owner the exclusive right to it, and the legal right to prevent others from either copying it or from 'passing off' goods or services as the owners of the mark. In the UK, it is the *use* of the trademark which gives its owner exclusivity of use and the goodwill or reputation associated with it.

Trademarks enhance the value of a company or its products by psychological means. Consumers associate the mark with one particular company and the quality or style of goods it produces. Sometimes it is the name of the company itself, such as 'Hoover'. In other instances, it is a particular product: how many people know the name of the company that produces 'Aspirin'?

In these two specific cases, the trademarks have become so familiar that the general public will often refer to any vacuum cleaner as a 'Hoover' or analgesic as 'Aspirin', even if the item is made by another firm. A company in this position can find themselves at a disadvantage, and can spend considerable time and money on suing other people who use their marks as generic terms. Indeed, if a trademark is deemed to have become generic, it is no longer considered to be a trademark and its registration has to be expunged.

Requirements for registration

A trademark should be:

- **distinctive**, but neither descriptive nor deceptive. Therefore avoid anything which either describes the actual product, or which might make it appear to be something that it isn't. 'Apricot' computers had their application for registration accepted, since the word did not have any obvious association with the product.
- **nothing**, even vaguely, **like any other mark** or name used by a competitor.

If you intend to export using the same mark, make sure that it is registrable overseas. And do make sure that the word doesn't mean something silly or obscene in the language concerned!

The rules governing registration seem somewhat arcane to the uninitiated, and expert advice should be sought. A commercial lawyer or patent agent is the best first contact. Not all marks are eligible for registration under the Trade Marks Act, so if you are starting business and are thinking of devising or adopting a mark, take advice before using it.

The process of registration

A patent agent will undertake a search of the Register for a modest fee, so it is neither difficult nor expensive to check your proposed mark for originality.

The process of registration, which takes up to two years, is as follows:

- **filing** the application at the Trade Marks Registry;
- a **search**, undertaken by the Registry to ensure that the proposed trademark is distinctive, is not deceptive and does not conflict with existing registered marks;
- **advertisement** in the Trade Marks Journal. Objectors have one month to register their complaint;
- **registration**, if no objections have been received.

UK registration does not provide protection in any other country, and it is necessary to apply for registration separately in each other country. The advent of the Single European Market means that it could well be advisable to register in other Member States. If you apply within six months of the UK application, you can claim priority from the date of the UK application.

Defending a trademark

You can defend your trademark against others who might try to use a similar mark, providing they are using it on goods or services which fall within the specification and class of your mark. The specifications and classes have to be clearly defined in the application for registration.

Remedies for infringement include:

- an injunction from the Court to prohibit further infringement;
- an order to destroy all the offending marks, whether on the goods themselves or on material advertising them. The goods themselves do not have to be destroyed;
- a proportion of the profits from the sale of the infringing goods or services, or damages based on lost profits due to lost sales.

As with copyright, there are also criminal sanctions available. Provided that fraudulent intent can be demonstrated, it is an offence to sell or deal in counterfeit goods which carry an infringing trademark.

Take it seriously!

Finally, the need to take intellectual property rights seriously must again be emphasized. The material presented in this Chapter is a digest of a very complex area of law, and an attempt has been made to simplify the subject (really!) and to make it directly relevant without being misleading. Nevertheless, it is appreciated that the average entrepreneur will still find it complex and legalistic. Understandably, you are more concerned with getting your idea into production and into the marketplace.

The problem is, that if you ignore the protection of the intellectual property which is yours, it is equivalent to leaving the factory door open on a Friday night: anyone can walk in and take your property away. And if you broadcast your idea widely before applying for patent protection, it is like leaving the factory door open and telling your competitors that you've left it open.

The other common failure is to apply for a patent to receive a priority date and then do nothing about it for nine months. There is then nowhere near enough time to exploit the invention or to file a second application with further information.

The key point, therefore, is to develop a rational *strategy* for exploitation. If you are starting a business yourself, it should be part of the Business Plan. If you are planning to sell the patent to a third party to exploit, then the work which needs to be done to file an application, the timing of the filing, the programme of work needed after receiving a priority date and the tactics and timing of licensing negotiations should be carefully thought through.

Today's consumer has an unprecedented choice. Products rapidly become obsolete, and products thus have ever decreasing lifetimes. If the principles on which the product operates are published too early, competition can render a product obsolete before it is launched. Securing a priority date is thus of little value unless you move rapidly to exploitation.

Appendix – Specimen confidentiality aggreement

SECRECY AGREEMENT

You or your company... is prepared to disclose to *ABC Company...* certain information developed by *you or your company...* relating to its business activities in *describe the general area of technology involved and the specific area of application...*

Since this information is held confidential by *you or your company...*, this disclosure to *ABC Company...* shall be made on the following basis:

1) *ABC Company...* shall maintain in confidence all information relating to *you or your company...* business activities in *describe the general area of technology involved and the specific area of application...* disclosed orally or in writing to *ABC Company...* by *you or your company...* and shall not, without prior written consent of *you or your company...*, disclose the information or use the information other than for the specific purpose noted above.

2) Each *ABC Company...* employee to whom any information, whether details or conclusions, is disclosed shall be informed of *ABC Company's...* obligations under this Agreement with respect to such information and shall have agreed in writing to hold such information confidential and not to disclose or use such information other than for the specific purpose of this Agreement.

3) *ABC Company...* shall make no copies of any documents provided by *you or your company...* under this Agreement other than for the use of the designated employees. Upon written request, *ABC Company...* shall return to *you or your company...* all documents provided by *you or your company...* and destroy all documents created by *ABC Company...* which incorporate any of the information provided by *you or your company...* or any conclusions reached by *ABC Company...* based on such information.

4) The obligations of paragraphs 1 and 2 above shall not apply to such information

a) which at the time of disclosure by *you or your company...* is already in the possession of *ABC Company*; or

b) which at the time of disclosure by *you or your company...* is, or thereafter becomes through no fault of *ABC Company...*, public knowledge; or

c) which after disclosure by *you or your company...* is lawfully received by *ABC Company...* from a third party who has the right to disclose the information to *ABC Company...*. To establish one of the above exceptions *ABC Company...*

shall provide written documentation of this disclosure in substantially the same degree of specificity as the disclosure was originally made by *your company*... .

5) This Agreement shall not be construed to grant to *ABC Company*... any licence or other rights, except as expressly set forth above.

6) The obligations of this agreement shall be in effect for a period of a *number of years, usually 3 to 5*... from the date on which this Agreement is executed between *ABC Company*... and *you or your* company...

7) This Agreement is personal to the parties hereto and the rights and obligations hereunder may not be assigned in whole or in part without the prior written consent of the other provided that either party shall be entitled without consent to assign the rights and obligations hereunder to any successor in title of its entire business in the products the subject of this Agreement.

If the foregoing terms are acceptable to *ABC Company*, please indicate agreement by signing and returning to *you or your company*... one copy of this Agreement.

SIGNED:

your signature	*your company*...	*date*
signature of authorised person	*ABC Company*...	*date*

Chapter 4
THE BUSINESS PLAN

Why have a business plan?

Why have a Business Plan? Because it enables you

(*a*) to plan to make a profit, and
(*b*) to persuade your potential investors that they too can make a profit if they are prepared to back you.

This Chapter will consider the essentials of Business Planning, with an emphasis on the new, technology-based project. It will concentrate on the planning necessary for start-up and raising finance, since this is the focus of this book. The next Chapter reviews the sources of finance for businesses and the problem of assembling an efficient package to help you carry out your Business Plan. Don't consider where and what investment to acquire until you know how much you want, when you want it and how much it will cost. In other words, tackle the business planning first.

A much more extensive planning exercise will be necessary during the growth phases of a company, and there is a companion volume in this series, 'A Business Plan' by Alan West, which contains more detail and advice.

■ PLAN TO MAKE A PROFIT

Unless you plan your business strategy carefully, there is no way of knowing whether the money put into your business would earn more in a building society. A Business Plan enables you to define targets against which you can assess the progress and success of your business.

It will also help you to put the various aspects of your business – product development, market strategy, property, personnel, and resources – together, into a coherent and credible picture. There is a favourite saying of the stressed businessman that 'I have been so busy fighting the alligators that I forgot that the original objective was to drain the swamp.' A properly devised Business Plan will at least allow you to keep the main objective firmly in view, and can go a long way to identifying the nature of each 'alligator' and when it is likely to give trouble.

There are two main areas of weakness which beset the new and growing business. The first is inadequate market information and the second is a lack of a proper financial structure and costings. A Business Plan can identify the precise areas of vulnerability, including management deficiencies, and greatly improve chances of success.

A Business Plan is not a panacea. Although it might identify the problem, it does not automatically provide all the answers. Also, the Business Plan should not be prepared, be used to assist in raising the capital and then be consigned to the archives. Of course circumstances will change and the business may be positively or adversely affected; but unless the 'map' is constantly available, there is no sure way of knowing the extent of the deviation. The Business Plan can provide a reference point for remedial action to be taken, or for a new opportunity to be seized.

■ PLAN TO PERSUADE YOUR INVESTORS TO BACK YOU

A Business Plan will be indispensable in securing the external finance necessary to launch a company, and in a technology-based business, where there is frequently a lengthy research and development phase before significant sales income is generated, this is particularly vital.

External investors in a business are naturally interested in the opportunity presented, but they are just as concerned to minimize their risks. Your Business Plan should be designed to convince them that you have:

- an identified market;
- a product or service to satisfy that market need;
- identified the financial needs to efficiently support the programme;
- a management team able to carry it out.

The Plan should be attractive, well-considered, positive but above all *realistic*. Although in a very real sense a Business Plan is a piece of marketing – selling your project to your potential backers – it is vital that your investors know that you are aware of the problems. They can spot rose-tinted spectacles a mile off.

Investors know that any business proposition carries some element of risk, and their overriding need in a Business Plan is to see that risk quantified and minimized. In a technology-based business there can be great rewards, but there are also substantial risks. What are they; how big are they; how can they be minimized?

Many investment organizations are not accustomed to dealing with or understanding new technology and its application; they are better at appreciating the opportunities offered by retail stores, light engineering and taxi services. This is perhaps understandable, but it means that any Business Plan presented to them which offers rich rewards, with little or no risk from a technology they find virtually incomprehensible, stands scant chance of receiving serious attention.

There are a large number of sources of investment finance available, from the high street bank to the venture capitalist (*see* Chapter 6). They vary considerably in the kind of project they are inclined to support, the risks they are prepared to run and the rewards they expect to make. Your local business advice centre, science park or innovation centre can advise you on the most appropriate for your particular project.

Before you start

■ SOME DON'TS

This book is intended to be positive, but there are some 'don'ts' associated with business planning, and perhaps it is best to get them out of the way first. Nevertheless, they are important.

Do not equate a Business Plan with a financial plan. The latter is only a part of the whole thing, and, indeed, the financial aspects can only be properly assessed once the other parameters are identified.

Do not just pass the whole exercise on to your accountant or a firm of management advisers. They may indeed be able to give invaluable help, but essentially the Plan must be *your* Plan. Even if you need help over much of its preparation, you must understand and agree with all the conclusions reached.

I can well remember sitting in on a meeting with an entrepreneur with a promising concept in water purification for medical use, who was seeking capital from a group of investors. When asked about a particular figure in his cash flow forecast, he replied: 'Oh, I don't understand these small details. I leave those to my accountant to work out for me.' Despite the fact that we carried on politely discussing the plan for another 30 minutes, it was clear to me that there was little point in continuing the meeting beyond that point.

Do not engage in false optimism. Be realistic, especially over sales levels in the early months. Many investment organizations will require financial forecasts at two or three levels of projected sales, and none will be satisfied unless you can justify your assumptions and targets. Also, false optimism in the planning stage will inevitably cause distress as soon as it is realized that the Plan is unrealistic.

On the other hand, do not overplay the difficulties. All business ventures involve a degree of risk. Your task is to identify them beforehand and show how they can be minimised.

Do not make the Plan indigestible. Keep the discussion fairly short (20 pages at most), and, if appropriate, put more comprehensive details in Appendices. A crisp, well-presented Summary page is essential. Even financiers are human, and conciseness and clarity will appeal to them.

■ THINK ABOUT THE PRODUCT

There are, of course, many businesses that provide a service rather than a product, and most of the following applies equally to technology-based service businesses.

First of all, *is* there a product? Is it still in your mind as a generalized concept, on the drawing board, part constructed or a working prototype? The danger is of drastically underestimating the cost and time needed for research and development.

Then, if there is a prototype, how long will it take and how much will it cost to get products on the shelf, ready for sale? A lot of blood, sweat and tears can be shed converting a perfectly good working prototype into production.

A common problem here is to use non-standard components in the prototype, and expect that they will be available from stock by the time production commences. So, are you sure *that that specialized chip will be available to the desired specification and reliability, and in sufficient numbers, by the time you need it? Even if the answer is yes, will its cost have been doubled?*

Do give thought to the back-up the product will need. Obviously, marketing and sales support will be required: the product will not sell itself. It will probably also need technical support for installation and maintenance.

The purchasers may well need some training in the use of the product, and they will certainly need good documentation. (How many times have software writers accompanied user-hostile programs with totally incomprehensible manuals?)

Think ahead, too, to the next phase of the business. Investors like to feel that ideas they support have a 'long mileage' in them. Many products have uses which were not originally considered for them. Could a small adaptation lead to a quite different market? If it does, do not discard your original ideas (unless it is clear that the new area offers much better prospects or a much lower cost), but consider the new market as a potential diversification for the future.

Also, all products have a finite lifetime, and, if the product is a good one, competitors will soon be arriving to reduce the original market share. Have you got a 'Mark Two' in mind, viz, a further development of your idea with another increment of originality in it?

■ THINK ABOUT THE PEOPLE

An American financier once said that investing in high-technology was like being a passenger in a Ferrari: 'You like to believe you're in for a wonderful time, but you sure hope the guy next to you can drive!' The success of a business ultimately depends on the people who run it, and it is of prime importance that the investor is given confidence in the management team involved.

I can well recall a meeting some years ago between the UK Science Parks Association and the British Venture Capital Association. The Science Park managers took the venture capitalists to task, comparing the UK scene with the much larger sums invested in new technology-based start-ups in the US. The reply was that they would dearly like to invest more in such ventures in the UK, since the quality of the products was high. The problem was, rather, with the quality of the management, *which at the time was generally not good. Much effort has gone into the improvement of management skills since that time, but we still have a long way to go.*

Chapter 1 p. 6 considered the qualities needed to be an entrepreneur, and it is important to think carefully about yourself and your management team when you are constructing your Business Plan. What is your real motivation, and that of your business partner or team? Do you want to run a company, or do you just want to see a commercial product come from your research laboratory? There is, of course, nothing wrong with the latter, but it might be better to let someone else with commercial experience and drive run the firm, so you can get on with inventing the next product.

You may be a technical genius, but much of running a small business is not technical. Have you got the ability, or even the patience, to successfully deal with staff, customers, market strategy and finance? Or you may be a successful sales manager in a large firm, and are contemplating moving out to start on your own, with a product whose market you know, but whose technical subtleties you don't. Can you deal with customers' technical queries, or develop new applications? These deficiencies will show up will they not, when you include your personal background in the Plan (*see* p. 92).

Think too about your partner(s). Do they just bring more of the same, or do they have complementary skills to your own? It is far better to have a mixture of attributes in a management team: even a certain amount of disagreement and 'creative tension' is preferable to unanimous agreement to proceed in the wrong direction.

Does the project depend critically on one man and his know-how? This can be dangerous. He may become ill or otherwise incapacitated. Have you a contingency to accommodate this? Also, such people can be brilliant and initiate commercial successes; but they can have 'butterfly' minds, and it can be difficult to get them to focus on one project for long enough to see it right through development into production.

Even if your team can cope adequately for now, will it be able to in a year or so, if the business expands in the way you hope it will? Starting a business is very difficult: managing a rapidly growing business is even more difficult (*see* Chapter 1 p. 9). What future gaps in management can you identify? Non-executive directors can be of great help here, if they can be chosen to provide a particular skill which may be needed. Venture capital organizations will often insist on non-executive Board representation to ensure that their investment is properly managed. Are these possibilities going to be given their due prominence in your Business Plan?

The elements of a business plan

■ A BRIEF SUMMARY

This is the Section to do *last*, once all the details have been put into perspective. It is one of the most important aspects, because the typical professional investor will generally not read the Plan any further if he doesn't like or doesn't understand the Summary. It should be crisp, readable and 'punchy' without being absurdly optimistic.

The Summary should extend for no more than 1-2 pages, and include:

- a brief background to your business or to your relevant experience. If it is not your business, who owns it?
- the essential features of your product.
- brief details of the market which you hope to sell to.

- basic information on location, premises required and number of people to be employed.
- how much money you want, when each instalment of cash input is required, and what level of return you expect.
- the principal objectives, with timescales. For instance:
  - what market share is anticipated?
  - when is profitability expected?

■ THE MANAGEMENT TEAM

A business stands or falls on three things: management, market and money. Perhaps the most important of all is the quality of the management team involved.

If you are the sole manager, then you will need to set out your suitability and track record, and also to undertake a brief but fairly critical self-appraisal of your overall abilities. Again, be realistic. It may be helpful to read Chapter 1 p. 11–12 before doing so.

Start by putting together career summaries of you and your management team; these can go into an Appendix, but you can extract from them the present track record and your future management strategy, with the roles to be filled. Does the team have the required complementary skills? Can the team handle a scheme of the size and rate of growth you anticipate, and in the competitive market you expect to enter?

Since we are dealing with technology-based companies, first ask if you really do have all the technical expertise you need. If you feel there are gaps, pinpoint them, and try to see how they can be covered. Is there a university or research institute nearby, with someone you can call on as a consultant, or equipment you can buy time on? If not, maybe there is a national resource you can tap, or a technical database which can be interrogated.

Then, are there other gaps in management skills? Have you the skill to manage the cashflow and the product development and the marketing strategy? How will you cover these areas? Perhaps not at management level, in which case you will begin to appreciate your personnel needs, which will be relevant for a later section of the Business Plan.

There are, of course, other ways of tapping into particular skills without enlarging the full-time management team. In the early days of, say, a small technical consultancy, it will not be necessary to go to the expense of employing a full-time financial manager; engaging a local firm of accountants to come in on a regular but part-time basis will be enough.

Use of other consultants and external advisers can be of value, but frequent use can be expensive, and in some instances may justify engaging an in-house expert. Non-Executive Directors can be very useful: they probably have a financial stake in the project's success, and, if properly selected, can provide vital experience without being on the permanent payroll.

Even if the skills of your management team are adequate for the present, will they be able to cope satisfactorily as the business grows? If you think that additional members of management may be necessary after a while, such as a financial or commercial manager, it is best to be aware of it at the beginning, and include it in your Business Plan.

Finally, how many other staff will be needed to develop and produce the product? *When* will they be needed? Will there be any special skills involved which may be difficult to recruit? If the product is highly technical, it may be necessary to pay a premium to get people with the right experience to work for you.

■ PRESENT STATUS

Many new ventures start with a pre-existing basis of activity or are diversifications or expansions of existing firms. If so, it is important to put the current state, reputation and 'track record' of the company in the Business Plan.

How long has the firm been in existence, and what have been the significant 'milestones' on the way? Where is it located, how many people does it employ and what kind of products does it make? Is your present market relevant to the one you will need to penetrate for your new venture? Be brief: if you think a more detailed background is germane to your Business Plan, put it in as an Appendix.

Finally, what are its present states of equity and finance? In other words, who owns the company, how much and from where has it borrowed, and is it under sound financial control? It will be appropriate to append the most recent set of audited accounts and perhaps the most recent management accounts, to illustrate and support your statements.

■ THE PRODUCT

The product (or service) is an obviously vital part of any new project. The nature and technical novelty must first be considered, and then the market, which is dealt with on p. 95. Remember that the initial assessment (and sometimes virtually all the assessment) of the project's investment potential will almost certainly be by people with little or no technical background. So the description of the product, its novelty and its principle of operation will have to be concisely and clearly described in non-technical terms.

Take great care over this. It *is* possible to explain the workings of highly complex systems in terms comprehensible to an intelligent layman. You only think you can't because you have been accustomed to mixing and working with other experts in the same field. I appreciate that I may be hoist by my own petard here, because this is what I am trying to achieve in Chapter 2! However, whether or not you get the support you need for your project may well depend on spending a considerable time in refining and clarifying the explanation of what your product is, how it works and what it does.

I have sometimes been faced by inventors who have not so much a product as a concept or technique which, it is claimed, has an immense range of applications. 'Don't worry about which product is going to be the first to exploit the idea; let's just construct a prototype to demonstrate the concept, and the requests to licence will come flooding in.' This is almost never true. In any case, investors will not be prepared to support a generalized concept. What they want is a marketable product with a price tag on it.

If your idea really does have a variety of good applications, obtain expert advice on patent protection (*see* Chapter 3) and go all out to produce the most appropriate first product based on the idea. Which application is chosen should be determined by the ease of achieving a market, and the time and cost of development.

It is important to explain clearly the advantages of the product over existing products. Since a highly technical product often finds its use in a highly technical area, this again must be carefully written for the non-scientist. However, do not be vague. Don't just say that your product is better – say *how* much better.

Evidence that the product will do what you claim for it should also be included. Describe its current state of development and include a realistic estimate of the time and cost to complete the development and feasibility studies. If there is already a working prototype, include a diagram or photograph, together with essential quantitative data on performance against the anticipated specification. Much of the detail can go into an Appendix.

Any Standards relevant to the product and its performance should be mentioned. These would include Safety Standards, legislative requirements and Performance Standards. Will it, for instance, need to undergo extended life trials or accelerated aging tests?

Explain also how the product is to be produced. Will it be manufactured entirely by your own company, or will you rely on it being partly or entirely made by subcontractors (*see* Chapter 1 p. 16)? Are some components or sub-assemblies difficult to make or obtain? These points should be considered, and any arrangements or agreements in principle with suppliers or subcontractors should be described.

Finally, if the product is novel, explain the 'intellectual property' situation pertaining to it. Is the concept patented or subject to copyright? Is it likely to infringe or circumvent someone else's patent? Who owns the relevant intellectual property rights? If you have a patent on the invention, what is its status (Provisional or Full)? How long will it be before refiling is due? In what countries has protection been obtained?.

■ THE MARKET

It doesn't matter how clever or original the product is, if it won't sell. The market for the product or service is all-important.

Perhaps understandably, inventors tend to become preoccupied with the technical content and novelty of their inventions. When inventors *do* attempt to assess the market for their ideas, they almost invariably don 'rose-tinted spectacles'. (Indeed, it would be wrong if they did not, since it is a poor inventor who is not also an enthusiastic 'champion' of his idea, and it is this enthusiasm which is often so vital to triumphing over the difficulties encountered in establishing and developing a new business.)

On the other hand, some of the best inventions come from people already working in the relevant field of application, and who from their own experience see the need for a product and the size of the market.

To counterbalance false optimism, someone with experience in market assessment should assist the inventor in the task of quantifying the potential market. The more objective and quantitative the assessment the better. It should cover

(*a*) the nature and size of the market and
(*b*) how the product will be promoted and sold to this market.

A market strategy

First consider the present state and size of the market. Is it small or large? Is it well served, with many potential competitors or is it totally open to your new product or service? Are there any new developments on the horizon which might quickly make the product obsolete? Is it static or growing? With many technology based products, the home (national) market may not be big enough to sustain the business, although it may be perfectly viable in an international market. This is one of the great potential advantages of the European Single Market.

Is your product one that will be readily accepted by existing users of comparable products? It is a substantial additional problem if the market has to first be educated before it can be sold the product. This is sometimes one of the difficulties to be faced when marketing a product based on novel technology, or technology which is novel to a new market (*see* Chapter 9 p. 202).

Who are the principal purchasers of your product – large corporations or individual purchasers? Is it a product which will be bought only once, or a consumable for which there will be a repeated demand? If you are in business already, are some of your present customers also potential buyers of the new product? It will certainly help if they are.

It is also important to have a realistic target for your share of the market. Remember that you have – or will have – competitors, and that they will not just stand by and see their market eroded. Their reaction to your product launch must be considered. Perhaps they will cut their prices or mount a promotion campaign: what will be your response?

This section of the Business Plan can be more easily compiled by undertaking the well-known 'SWOT' Analysis: Strengths, Weaknesses, Opportunities and Threats. Make a list of the factors affecting each of these, and incorporate them into an overall market assessment.

A promotion and sales strategy

Promotion and sales are essential aspects of marketing, but do not imagine that the two are identical. Promotion is the activity of raising public awareness (or, better, potential customers' awareness) of the product and its positive attributes, so that a desire to buy is created. Selling is the conversion of that desire into positive action. It is vital to consider both aspects.

□ The distinction between promotion and sales is well illustrated by the partially deficient market strategy used for a very good series of open-learning training programmes for small businesses.

The series was attractively produced, received excellent reviews in the relevant specialist press, was glossily promoted and the exhibition stand for the training programmes won prizes for its quality at national exhibitions. Thus, the *promotional* strategy involved was very good. There was good evidence that the product was well known to training specialists and enjoyed a good image.

Despite this, the programmes did not sell. It was quickly realized that a positive sales strategy was needed, and a good telephone sales person was employed to capitalize on the effective promotion. It was not long before sales dramatically improved, and the series eventually became very successful. □

The Business Plan should thus contain a summary of the promotion campaign designed to create awareness of the product within the target market and to stimulate demand. This may be done by creating an appropriate corporate image or simply by making the product attractive.

Pricing strategy also needs considerable thought. For instance, if your product is, for some reason, much cheaper to make than competitive products, do not necessarily price it at the lowest possible. It may be better to undercut the opposition, but still leave some further room for price reduction should the competition try to undercut *you*. Chapter 7 p. 174 deals with pricing in more detail.

Some of the most successful products are sold not just on the qualities of the product itself, but on its ready availability and product support. So, what are your tactics for distributing your product? Will it be sold solely from your premises or will you use dealers or distributors? If the latter, are they reliable and will they cover adequately the geographical area needed? What after-sales service will the product require? Could you beat the opposition by offering a better service, or might it prove too expensive to maintain?

Provision of such customer advice and support should be part of the overall marketing strategy. This is particularly important in technology-based businesses. If your customers feel they can rely on a rapid response to their queries or difficulties they will come to trust you and are more likely to buy from you again.

In summary, the market for the product and how it is planned to address it is the core of a Business Plan. All the sophisticated computer generated cash-flow analyses in the Financial Section will be of little value if the market strategy is fundamentally wrong. It is *vital* to take the greatest care and thought over this section of the Plan.

Financial projections

The financial needs and projections for a Business Plan should flow out of the information and assumptions made in the other Sections. This may seem obvious, but there are too many Plans whose financial sections seem divorced from anything that has gone before. Employing an accountant to help can be an immense aid to sound financial forecasting, but make sure he uses *your* market assessment and sales assumptions and does not invent some entirely of his own.

Do not be scared of asking for enough money. Trying to make do with as little as possible can rapidly lead to cash flow problems. It has been estimated that even for a tightly financially controlled manufacturing company, supporting an increased turnover of £100,000 requires an injection of around £25,000-£30,000 to finance it.

Before looking at Financial Plans in more detail, there are a few pieces of general advice which might prove useful:

- get someone to check and question your assumptions. Use your accountant or a friend who is prepared to be constructively critical. In a business with an innovative product it may be quite difficult to establish initial forecasts of, say, development costs or sales. So include adequate funds to accommodate delays in and additional costs of research and development, initial production problems, etc.
- do not make attractive assumptions 'just to make it come out right.' If you do, it will probably mean that the business is inherently unprofitable or that too little cash will be requested. It will also not be long before the external investors begin to question your overall forecasting skills. One hears a lot about 'bottom line' credibility. 'Top line' credibility is even more important: the initial assumptions *must* be realistic.
- use a computer to prepare the forecasts. There is much excellent financial software on the market, and your accountant, local business advice or innovation centre will be glad to advise. There is no need to go to extremely expensive software: for instance, Supercalc 5 only costs around £80 at present, and offers very good value for money. Alternatively, integrated suites of software like Lotus Works, MS Works or Framework, incorporating a simple word processor, database, graphics and spreadsheet can be purchased for £150-£200. Use of a computer makes it possible to carry out any number of 'what if?' forecasts. Quite apart from the benefit this can give you, professional investors frequently ask for financial projections at two or three levels of sales.

■ PROFIT AND CASH FLOW FORECASTS

The financial projections should include both a profit and cash flow forecast, together with a balance sheet projection at each year-end.

The first, because if there is little or no profit in the scheme no external investor will want to put money into it – and neither should you!

The cash flow forecast is vital because it is only too easy for a highly successful expanding business, with full order books, simply to run out of money and thus fail. Let us now examine both these projections in more detail.

The profit projection

This should be a monthly forecast for the first year, a monthly or quarterly forecast for the second year and a quarterly forecast for the next two or three years. It should be attempted at two or three different levels of sales (and thus of variable costs), usually

(*a*) at the level reasonably anticipated,
(*b*) at a pessimistic level (the 'worst case') and
(*c*) at the most optimistic level.

A simplified example is shown in Table 4.1. It is the kind of initial projection which would be made right at the start, before any assessment of the required investment is available; indeed, it is this sort of projection which will produce an indication of the cash needed to start up. It would normally include all income expected from sales, grants, loans and other investments, and all committed expenditure. It would also incorporate charges for finance, loan, leasing and hire-purchase payments, and any legal and audit fees. The objective of the forecast is to show that the assumptions made lead to a satisfactory level of profit within a reasonable time.

The forecast in Table 4.1 indicates that the maximum investment required to get the business off the ground is just over £30,000, and that this could be repaid by the end of the first year. The investment need in fact only exceed £30,000 for one month (Month 3), and the Table shows that the firm is profitable after Month 11. The cash flow picture, however, predicts rather a different story (*see* below).

It would seem apparent that a firm involved in fish and chips should become profitable sooner than one developing an innovative application of silicon chips, where substantial time and investment in equipment, research and development might be necessary before the realisation of substantial sales.

Investing organizations and their shareholders, however, demand a return on their investment as soon as possible. This obsession in this country with 'short-termism' has been the subject of much discussion, and there is evidence that the UK has missed several first class opportunities to profit from innovation by being unwilling to wait an extra year or so to realize the return on the investment.

MONTH	1	2	3	4	5	6	7	8	9	10	11	12	TOTAL
INCOME													
Sales	2000	5000	8000	10000	10000	14000	15000	15000	15000	15000	18000	18000	145000
VAT at 17.5% of sales	350	875	1400	1750	1750	2450	2625	2625	2625	2625	3150	3150	25375
TOTAL INCOME	2350	5875	9400	11750	11750	16450	17625	17625	17625	17625	21150	21150	170375
EXPENDITURE													
Salaries plus PAYE and Ins	3600	3600	3600	3600	4800	4800	4800	5520	5520	6600	6600	6600	59640
Rent, rates	880	880	880	880	880	880	880	880	880	880	880	880	10560
Purchases plus VAT	10000	1000	3000	3200	3200	4000	4200	4200	4200	4200	4500	4500	50200
Travel	400	300	300	300	300	300	300	300	300	400	300	300	3800
Light and heat	55	55	55	55	55	55	55	55	55	55	55	55	660
Telephone, fax	120	100	100	100	100	100	90	90	110	90	100	110	1210
Postage, stationery	300	80	120	60	70	90	80	80	130	100	90	90	1290
Marketing	3000	1500	2500	1500	1000	1500	1000	1000	2000	1500	1000	1000	18500
Capital equip.	10000	0	0	0	0	0	0	0	0	0	0	0	10000
Insurance	2000	0	0	0	0	0	0	0	0	0	0	0	2000
Depreciation												3300	3300
TOTAL EXPENDITURE	30355	7515	10555	9695	10405	11725	11405	12125	13195	13825	13525	16835	161160
MONTHLY PROFIT	−28005	−1640	−1155	2055	1345	4725	6220	5500	4430	3800	7625	4315	
CUMULATIVE PROFIT	−28005	−29645	−30800	−28745	−27400	−22675	−16455	−10955	−6525	−2725	4900	9215	9215

Table 4.1 Profit forecast.

There is some evidence now that investors are beginning to appreciate that investing in technology-based enterprise can be highly *profitable, but that it often takes some time. Equally, however, the Business Plans and the assumptions in them must be credible, and the management teams involved must be committed to achieving their objectives and timescales.*

The cash flow forecast

This should be on the same time basis as the profit forecast. Table 4.2 shows the same Profit Forecast of Table 4.1 translated into cash flow terms. They look deceptively similar in format and have a similar year-end outcome, but have *significantly* different implications from the viewpoint of the cash required.

Before considering the example in more detail, it is worth making a few general points, since, despite the simplicity of the principles involved, failure to control cash flow is probably the biggest single cause of business failure of high-growth companies.

The essential difference is this. If you buy something for £1,000 and sell it the same day for £1,100, then the budget forecast will indicate a profit of £100. However, if you paid the £1,000 in cash on the day the item was bought but you are told by your purchaser that you will not be paid for a week, then for that week your cash flow is £1,000 'in the red'. If you are a business you would be insolvent unless you had arranged to borrow the £1,000 for a week from somewhere.

On this basis, increased sales only makes things worse, since your bank balance only changes when money is paid in or out: it does not depend on the day when the purchase or sale was made.

The essential difference is therefore between the contractual *commitment* to a sale or purchase and the time of the *actual transfer* of the cash involved.

The cash flow forecast is built on a large number of factors: prices, the timing of purchases and sales, the scale and timing of other income and expenditure. Table 4.3 includes the most common factors, but is not intended to be exhaustive.

It may at first sight seem surprising that so many businesses fold through a failure to appreciate such apparently simple matters, but, of course, cash flow prediction for a business can be quite complex. Until you become accustomed to the skills involved, it is important to seek expert advice. It has to include, for instance, monthly or quarterly repayment of loans or interest, leasing charges, rents, insurance premiums, VAT, PAYE, etc., besides the expected purchases of consumables and salary bills.

MONTH	1	2	3	4	5	6	7	8	9	10	11	12	TOTAL
INCOME													
Sales receipts (inc. VAT)	0	705	3173	6580	9753	11515	13160	16333	17508	17625	17625	18683	132658
TOTAL INCOME	0	705	3173	6580	9753	11515	13160	16333	17508	17625	17625	18683	132658
OUTGOINGS													
Salaries	3000	3000	3000	3000	4000	4000	4000	4600	4600	5500	5500	5500	49700
Rent, rates	1950	0	0	1950	0	0	1950	0	0	1950	0	0	7800
Purchases	0	9400	3290	3055	3713	3760	4512	4888	4935	4935	4935	5217	52640
Travel	0	400	300	300	300	300	300	300	300	300	400	300	3500
Light and heat	0	0	110	0	0	165	0	0	165	0	0	165	605
Telephone, fax	0	0	220	0	0	300	0	0	280	0	0	300	1100
Postage, stationery	200	60	95	45	100	90	75	85	90	90	70	90	1090
Marketing	0	2400	1800	2300	1700	1100	1400	1100	1000	1800	1600	1100	17300
Capital equip.	0	8000	2000	0	0	0	0	0	0	0	0	0	10000
Insurance	2000	0	0	0	0	0	0	0	0	0	0	0	2000
VAT @ 17.5%	0	0	1225	0	0	4900	0	0	7700	0	0	8400	22225
PAYE, Nat. Ins.	0	600	600	600	600	800	800	800	920	920	1100	1100	8840
TOTAL EXPENDITURE	7150	20860	9640	6300	6413	11415	7087	7173	15390	8045	8105	16672	124250
MONTHLY PROFIT	−7150	−20155	−6468	280	3340	100	6073	9160	2118	9580	9520	2011	
CUMULATIVE PROFIT	−7150	−27305	−33773	−33493	−30153	−30053	−23980	−14821	−12703	−3123	6397	8408	8408

Table 4.2 Cash flow forecast.

RESEARCH AND DEVELOPMENT

- time taken
- cost involved
- grants for development, feasibility study
- protection of invention e.g., cost of patenting

MARKETING

- cost of market survey and strategy
- pricing strategy
- promotion and advertising

FIXED ASSETS

- premises, rental method, rates, property maintenance
- other capital costs and method of payment

FINANCIAL COSTS

- loans, interest rates (and projected future rate)
- costs of administration
- effect of inflation on purchases and sales
- professional fees
- taxes: PAYE, VAT, Corporation Tax

PURCHASES

- cost, credit arrangements and method of payment
- availability of components and discounts for quantity

PRODUCTION

- premises: rental, rates and method of payment
- staff and labour costs
- sales compared with production
- quality assurance: level of faulty products

DISTRIBUTION AND SALES

- transport costs
- sales or distribution agency arrangements and fees

CREDIT CONTROL

- method of payment
- actual length of credit time
- allowance for slow payers and bad debts
- legal costs involved in credit control
- administrative costs involved
- bank clearance times

Table 4.3 Some factors involved in cash flow forecasting.

On the income side, in addition to sales income (which should be realized 30-40 days after the sale is agreed, depending on the terms of the sales contract), a cash flow forecast should include when and how much money is to be injected into the business. It should also include any government grants.

The government seems to be among the last to appreciate the disastrous effects of cash flow on a growing, small business. I have witnessed the generation of several managerial ulcers by substantial grants being agreed but not being paid at the appropriate time.

In addition, not all your debtors should be expected to pay up within the contractual time. Some customers are exceedingly slow to pay and a few will not pay at all.

Now look at the two forecasts in Tables 4.1 and 4.2 for the same operation, to illustrate the effect of cash flow on a business. The principal assumptions made for the cash flow were:

(*a*) that 30% of invoiced sales were paid to the company within 30 days, 60% between 30-60 days and the remaining 10% by 90 days;
(*b*) that the company paid 80% of purchase invoices within 30 days and the rest within 60 days;
(*c*) that a quarter's rent and rates were due in advance;
(*d*) that most of the other bills were paid quarterly.

Fig. 4.1 compares the cumulative forecasts graphically, and it is clear that the predicted maximum negative budget and cash flows occur in the same month (Month 3) but that the cash flow remains at around −£30,000 for a further four months, whereas the budget figure rises much more rapidly. The true maximum cash requirement is nearer to £35,000 and, in addition, the cash deficit in Month 9 is more than twice that of the profit forecast.

The differences can be much more drastic than this. For instance when a customer delays paying a large bill, or if sales take a month or two longer to pick up.

This can be true frequently for technology-based businesses, whose owners often have the impression that the market will take to their product quicker than it does in reality. Fig. 4.2 shows the same profit forecast as Figure 4.1 and Table 4.1, but with the income from sales delayed. The following changes were made in the

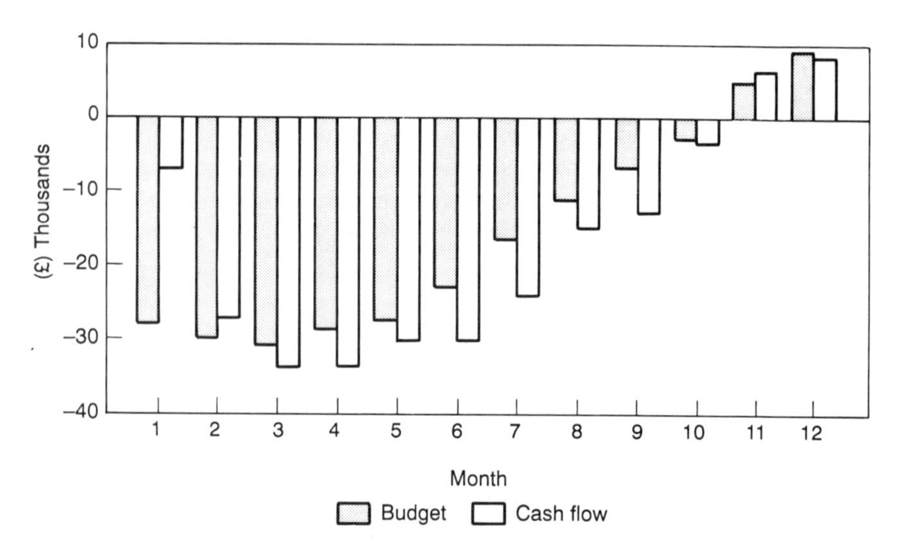

Fig. 4.1 Plot of budget and cash flow from Tables 4.1 and 4.2

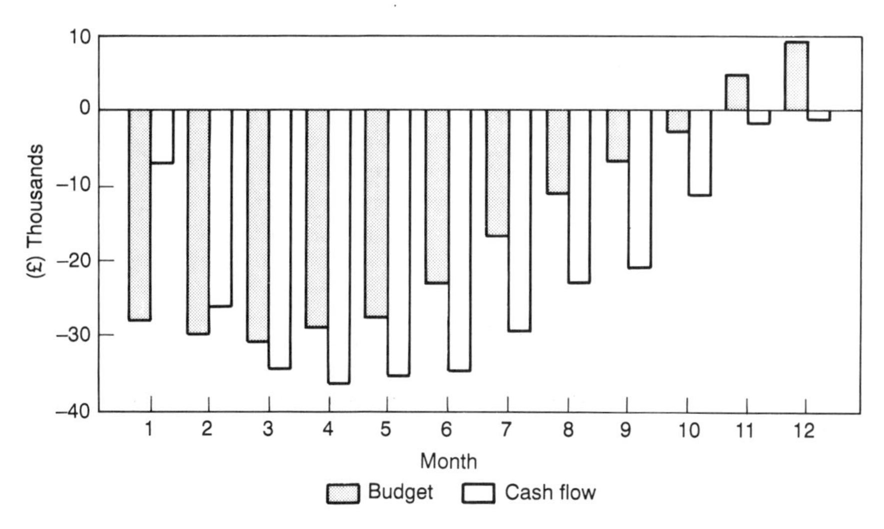

Fig. 4.2 Plot of budget and cash flow with deferred sales income

assumptions:

(*a*) that only 10% (and not 30%) of invoiced sales were paid to the company within 30 days, 30% instead of 60% between 30-60 days and the remaining 60% still within 90 days;

(*b*) that the company realised that the invoices were either slow in going out or in being paid, and it therefore held back slightly on payment of its purchases. It thus only paid 70% and not 80% within 30 days and the remaining 30% within 60 days.

The effect on cash flow is significant. The maximum cash requirement rises to nearly £40,000 and the loss stays below £30,000 for four months. Indeed, the company is not out of the red even by the end of the year. This example, albeit simple, illustrates the need not only to predict cash flow, but also to closely monitor it month by month, or even week by week. Chapter 7 p. 169 emphasizes this point in some detail, in the context of financial management of an operating business.

The balance sheet

The budget and cash flow projections should enable you to produce forecast Balance Sheets for the company after each year. The Balance Sheet is really a 'snapshot' of the firm's financial position at a given date, and examples are given in Table 4.4. They show how the funding, provided by the shareholders and retained profits, has

BALANCE SHEET AT START-UP		BALANCE SHEET AT LATER DATE	
Assets		Assets	
Fixed Assets		*Fixed Assets*	
Premises	–	Premises	–
Plant and Machinery	18250	Plant and Machinery	38240
Fixtures and Fittings	400	Fixtures and Fittings	2298
Vehicles	9200	Vehicles	16202
Current Assets		*Current Assets*	
Stock	2480	Stock	22361
Sundry debtors	–	Sundry debtors	21017
Cash at bank	24700	Cash at bank	6824
Cash in hand	550	Cash in hand	624
	55580		107566
Capital and liabilities		Capital and liabilities	
Capital		*Capital*	
Capital input: self	17580	Capital input: at start	37580
Capital input: investors	20000	Capital input: additional	20000
Net profit during year	–	Net profit during year	12448
less Drawings	–	less Drawings	–21358
Long-term Liabilities		*Long-term Liabilities*	
Mortgage	–	Mortgage	–
Bank Loan	18000	Bank Loan	26000
Current Liabilities		*Current Liabilities*	
Sundry creditors	–	Sundry creditors	32896
Bank overdraft	–	Bank overdraft	–
	55580		107566

Table 4.4

been used to fund the fixed assets and working capital (*see* Chapter 7 p. 176).

The Balance Sheet at start-up is a relatively simple document. The assets are likely to be no more than the initial purchases of equipment, stock and any vehicles bought by the company, together with the cash assets, most of which will probably be in the bank. In the simple example, the liabilities are the capital put in by the owner plus that of any external investors who have been persuaded to take equity, plus an additional loan from the bank.

The profit and cash flow predictions enable a forecast of the Balance Sheet to be made at some point in the future. The actual operation of the company brings additional assets and liabilities, but its overall value has increased. One additional asset is the money owed to the company by its debtors (i.e., its customers), but this is somewhat countered by the money now owed *by* the company to its creditors. Notice too how the additional activity of the firm has been funded by additional investment and loans, together with profit from trading.

The Balance Sheet is used to estimate a number of performance indicators, and these are discussed in Chapter 7 p. 180.

■ PUTTING IT TOGETHER

Once the draft Plan has been written, you and your management team – plus, probably, your accountant or financial adviser – should meet to review the assumptions and consider contingencies.

Review the basics: the management, the market, and the financial requirements. Consider the timescales: for instance, is the product development time enough? Are the promotion and sales properly sequenced? Have we got the right skills to achieve the objectives?

Do some *sensitivity analyses*. This is just a fancy term for 'what if' analysis. 'What if' sales were only 80% of the predicted total over the first year; 'what if' interest rates rise 2%; 'what if' it takes six weeks longer to get into production? Your Plan might prove quite robust against some changes, but might be very sensitive to others: this is where the term comes from.

Is your Plan therefore robust enough to stand up against possible additional adverse factors? Equally, 'what if' orders come flooding in and we need to buy in much more stock at an early date: will the cash flow stand it?

All these points will serve to sharpen up the objectives in your mind as well as making a better Plan. Such thorough analysis will also stand you in good stead when you come to argue the case for investment with the banks or venture capitalists. The fact that it is *your* Plan and you know it inside out will stand out a mile, and is bound to act in your favour.

Finally, make a decent presentation of it. There is no need for pages of multi-coloured graphs, illustrating the power of the graphics software at your disposal. But there is good reason to use attractive, relevant diagrams and charts to enhance your main themes. Uncluttered layout can improve readability substantially.

Conclusion

There are four principal reasons for business failure, all of which can stem from inadequate management skills or foresight:

- failure to control cash flow;
- insufficient funding to meet the objectives;
- poor pricing policy;
- lack of clear objectives.

A Business Plan forces some disciplined thinking on these aspects. It may be a somewhat painful experience to go through, but it is much less painful than a business failure. There is clear evidence that the long-term success of a business is dependent on the amount of planning carried out.

Chapter 5

SOURCES OF FINANCE: AWARDS, GRANTS AND SPECIAL LOANS

Finance for technology-based businesses

Securing the finance for a technology-based business usually means not only finding the money for production, marketing and sales, but also for research and development. However, it is all part of the project cost, and it is, or should be, all part of the Business Plan.

This is the key starting point. The entrepreneur attempting to establish a technology-based business faces a difficult task in raising finance, and the production of a realistic, costed Business Plan is *absolutely vital* to its success. Do read Chapter 4, and take all necessary advice to ensure a sound Business Plan.

Taking advice is very important. If possible, get an adviser who is not employed by one of the financing institutions themselves. The following represents a range of possible sources of advice. Which you use is up to you, and your choice will depend on who you know, your past experience, your needs and your knowledge of the local scene and its advice network.

- Accountants
- Solicitors
- Enterprise Agencies
- Science Parks
- Small Firms Advisory Service
- DTI Offices
- (Small) Business Clubs
- Chambers of Commerce
- Bank Managers
- Guidebooks/Directories
- Innovation Centres
- Management Consultants
- Business Colleagues
- Local Authorities
- Libraries
- Research Associations

Also be sure that the advisers themselves have experience in assisting the sort of business you want to start: an Enterprise Agency which is first class at helping typical small manufacturers may not have the necessary know-how when it comes to advising on a business based on technology, with a cash hungry R&D phase.

■ DIFFICULTIES INVOLVED

Technology based businesses, or, rather, entrepreneurs with such projects, are still at a disadvantage when it comes to raising money.

The reason often given is that technical people are not good managers, and that it is the management team which makes or breaks a project. This is undoubtedly true, and much work needs to be done in assisting entrepreneurs to enhance their management skills. However, in my experience, the same is almost as true of an equal number of entrepreneurs with *non-technical* business ideas. In reality, it seems often to be an excuse on the part of the financial institutions, who haven't got much of a clue about technology and are too hesitant to do anything about it.

There is still a very long way to go, but there is a slowly increasing realization by the banks and financial organizations that the special features of technology-based projects need proper treatment. There is also a recognition that getting a stake in some of these projects at an early stage could be very profitable in the long term. Although the risk is relatively higher, the initial outlay is still small.

Complex products and principles

One can perhaps understand the reluctance of lenders to invest in technology-based ideas. Such projects frequently involve complex products whose function is obscure, and even if the product is relatively mundane, it may be based on technically subtle principles which are difficult for the layman to grasp. Typical bank managers have great skills and personal understanding (and are a much maligned breed!), but are usually more accustomed to assessing the prospects for dress shops or small printing works.

Venture capitalists often exhibit a similar aversion to anything technological. They claim that investment in property or retail services presents their shareholders with better prospects, despite some spectacularly bad decisions in these areas in the late 1980s.

These views cannot simply be dismissed as prejudice. If you are seeking investment you need to deal with these people and deal with them on their terms. It is important to understand how they work and think, which is part of the reason for this Chapter. One thing, however, is for sure: you will (rightly!) not get anywhere without a sound Business Plan.

Long-term projects

Many knowledge-based ideas need a lengthy and costly research and development phase before any kind of product is produced. Indeed, some such businesses are totally research oriented, and are never intended to produce or market a product themselves (*see* Chapter 1 p. 14).

Also, a totally new concept will not necessarily find the identified market ready and waiting to accept it: there sometimes needs to be a period of market *education* to enable the market need to be created. (Of course, for a first rate product, it is worth waiting for, since, once accepted, it can rapidly achieve almost monopoly status).

The financial world is not currently accustomed to waiting a long time – four to seven years – for a return on investment, however large the ultimate size of the prize. Investors have become obsessed with 'short-termism', and whilst it would be nonsense to expect every investment made to be a very long-term one, it would make strategic economic sense if a much higher proportion of them were.

No track record

As indicated in the previous section, the technology-based business may be attempting to develop a totally unique product or service. This is both a plus and a minus. The plus is obvious: you have the market to yourself. The minus is really for the same reason: you will not be able to say with any certainty what the market will be or how fast it will grow, and hence it will be difficult to forecast sales accurately.

There is also the matter of your own, or your management's track record. All business ventures involve risk, and the financial institutions recognize this. However, they want to see the risk minimized, and quantified. They will look very carefully at the management team, to make sure that they have the right blend of skills and experience.

Lack of assets

There is frequently little the technology-based entrepreneur can offer the investors as security for their investment, other than his inventiveness and experience. More conventional businesses have tangible assets which can be offered as security against the failure of the business.

In another sense, of course, the innovator's inventiveness and experience is the key asset, without which the Business Plan would be unfeasible: but this particular asset is literally priceless at the time of first seeking financial support, and will have no resale value at a later date if the business is unsuccessful.

Competition

Many technology businesses are subject to fierce competition. Even those which initially have the market to themselves may soon find that their product is obsolete, having been superseded by more advanced technology from elsewhere, often from another country.

'Elsewhere' is a very big place in the context of technology. If the market is big or lucrative enough, the competition will be along, and sooner rather than later. Even if your idea is well protected by patents, it is still likely that attempts will be made to circumvent them by some means.

What this implies is that you, too, must look for overseas markets: the creation of the Single European Market can be used to your advantage just as much as to that of your competitors. It also means that you should try to maintain your inventiveness, and develop an equally revolutionary 'Mark 2' product before your competitors get around to copying 'Mark 1'.

Types of finance

It is important to ask for enough money to achieve the objectives of the Business Plan. One of the more common causes of the failure of new businesses stems from their managers trying to make do with insufficient finance, which then leads to cash flow problems.

Basically, the types of finance usually envisaged are:

- your own money, or that of your relatives and friends;
- grants and awards, which can be very useful, but do not normally constitute the main element of the finance;
- loans, which means borrowing some money from the bank or some other reputable organization, and then paying it back in instalments, with interest;
- equity finance, which means selling part of your company to someone else or to another organization in exchange for their cash input. The investor hopes to gain by seeing the value of his investment grow as the value of the company grows. The speed with which he gets his money back, and whether he gets periodic interest or not, can vary.

Putting together a financial package necessary to start or diversify a business is normally an iterative process. It should start with an assessment of the maximum amount of money needed, and of when the major cash injections are required. Initial negotiations with sources of finance and with advisers on grants and loans is the next step. This will lead to a first attempt at an overall package, which will probably include finance from a variety of sources. This is then taken through a series of refinements by negotiation, until both the entrepreneur and the investors find it acceptable.

It is a matter of making sure you can get a package you can 'live with'. Sometimes, negotiations will progress steadily towards a satisfactory deal, but on other occasions the only thing to do is to jettison the original strategy and start all over again, with perhaps a different strategy or a different set of potential investors. It can be frustrating and irksome, and, although there are no hard and fast rules or strict mechanisms, the following ways of approach may be useful.

■ FUNDING NEEDS

There are various ways of listing the sources of finance available, e.g., bank loans, grants, equity finance, research awards, joint venturing.

One useful rule is to try to match the size and time scale of the finance to the purpose for which it is intended. Do not, for instance, use a short-term overdraft to buy plant or equipment. The major types of finance, possible sources and appropriate purposes are listed below.

Short-term Funding

Duration: up to three years.

Sources: self, family, friends, retained profits, the clearing banks (as overdraft or short term loan), finance houses, leasing or hire-purchase companies.

Purpose: working capital over the short term, or bridging finance.

Medium-term Finance

Duration: medium term, say three to eight years.

Sources: self, family, friends, retained profits, banks, development corporations, finance houses, leasing companies, government and EC loans and grants.

Purpose: Longer-term research and development, medium-term assets, machinery.

Equity Finance

Duration: long term.

Sources: self, family, friends, retained profits, private investors, trusts, development corporations, merchant banks, venture capital funds.

Purpose: research and development, long-term capital, forming the permanent financial base of the company.

Long-term Finance

Duration: Long term, say over eight years.

Sources: self, family, friends, retained profits, banks, development corporations, finance houses, leasing companies, government and EC loans and grants, insurance companies, pension funds.

Purpose: Long-term assets such as land and buildings. Usually associated with substantial corporate development plans.

This is all pretty heady stuff, and for the technical entrepreneur with little financial experience it can seem rather daunting. There is an alternative approach which is entirely pragmatic, and is based on the entrepreneur's need to acquire sufficient finance to establish his or her business at as low a cost as feasible. So, with apologies to sophisticated colleagues in the financial world, one can classify finance as:

- your own money;
- free money;
- cheap money; and
- expensive money.

It is important for the entrepreneur to realize that money is a commodity, just like sugar or petrol. So if you are contemplating raising finance, look for the 'loss-leaders', special offers or deferred payments.

PUTTING THE PACKAGE TOGETHER

Before seeking finance, plan out your needs, so that you know how much you want, when and for what purpose. First, how much of your own resources might you be able to reasonably afford? Then, consider the free money: the grants and awards which might be relevant. Next, see if there are any low-interest loans which are applicable, or whether any scheme of subsidised help might be available, e.g., for a marketing plan or to undertake some development work.

It is equally important to use the money for the purpose for which it is intended. I have passed businesses many times and seen gleaming new BMWs standing outside, knowing full well that the company to which they belong is having difficulty in financing its working capital.

Bank loans and equity, the more expensive forms of finance, will probably still form the bulk of the capital required. This is not surprising, since banks cannot live by special offers any more than Tesco can. However, just as Tesco's and Sainsbury's prices will not be identical, neither will the cost of borrowing from the banks. So do shop around, to see if one source will treat your specific proposal more sympathetically. On the other hand, do not lose a manager who you have known and respected for years for the sake of an insignificant improvement in terms.

It is usually inadvisable to seek finance sequentially. It can take a long time. For instance, a maximum overdraft or loan facility can be agreed while the results of a grant application are pending.

Many entrepreneurs just stick to one source of finance, usually the high street bank where they have had a personal account for many years. This is not always the best way, and I have seen people whose personal finances have been stretched to the limit, when, had they taken the trouble to find out, they were eligible for a substantial grant to cover certain aspects of their expenditure.

In some instances, it makes better sense for some of the costs, and eventual profits, to be shared by corporate venturing, that is, by allowing another trading company to take equity and/or to input its particular relevant skills into your company. Maybe you have invented something, and can make it in sufficient quantities, but you have not the organization to market the product to its potential purchasers. My Innovation Centre, for instance, was confronted by two entrepreneurs who had invented a novel device for domestic central heating systems. They could make it in quantity, but had no way of attacking the national market which would assure the success of their venture. We were able to put them in touch with a large 'DIY' chain, who promptly offered to market and distribute their invention nationally, for a share of the sales revenue.

This is a very simple example of corporate venturing just to illustrate the concept. It embraces everything from straightforward collaborations like this, right through to purchase of equity and Board representation. It can include agreements on various aspects of the business, from research and development, to manufacturing and marketing.

Technology transfer and licensing may be regarded as being at one extreme of the total joint venture arrangement possibilities. The total responsibility for manufacture, marketing and selling is taken on by another firm. This is dealt with in Chapter 3, as part of the discussion on Intellectual Property Rights.

This Chapter now goes on to consider the sources and availability of 'free money' and 'cheap money'; the subsequent Chapter considers bank loans and venture finance, including corporate venturing, which are more expensive.

Personal finances

Any financial package must start by including some of your own money, although it is appreciated that this is, to you, neither free nor cheap.

You should think very carefully about the amount you are prepared to risk. It is strongly suggested that you discuss this in detail with an independent and trusted adviser. If you are unprepared to invest any of your own money, the lending institutions will feel, with some justification, that you have no real faith in your project and will doubtless be guided by this indication themselves.

It is also probable that the banks will want additional security for their investment, although there are now some 'seedcorn' schemes offering unsecured loans. Once again, think very hard before 'putting your house on the line'. It *may* be the best decision you ever make, and enable you to secure the capital you need to initiate a business which will make you very rich. On the other hand, it could spell unhappiness and ruin to you and your family.

Whilst on the subject of personal finances, it is vital for the entrepreneur not to give up any current job until it becomes absolutely essential. I have seen entrepreneurs – and their families – suffer appalling stress and hardship through ignoring this advice. So, if you have a good business idea and are in a job, work at your idea in the evenings and weekends and at every spare moment, but keep your job until the very last moment. You will still suffer stress, but not nearly so much as you would if you cannot pay the mortgage.

Free money

One Government Minister for Small Businesses once confided to me that 'My Government runs 99 grant schemes to help small businesses, but I only know about 20 of them in any detail'.

Since any relevant grant will be welcome in principle, it is important for the entrepreneur to find an adviser who not only knows about *all* relevant grant schemes but is able to advise and select those which are appropriate in each instance. Of course, there is no totally free money: all grants have various conditions and 'strings attached'.

These conditions are sometimes quite complex, and much time could be spent unravelling the small print only to find that the scheme is inapplicable to the project. Your local Enterprise Agency, Small Firms Advisory Service, DTI Office or Innovation Centre are normally quite experienced at advising grant-seekers. The National Westminster Bank also has a free service providing a list of grants available.

Do try to find someone who knows his or her way around the grant scene. Jim Cuerden, my own grant adviser at Newtech, has assisted in the acquisition of over £1 million in Government Grants for local small firms each year since 1987 – and in many instances the entrepreneur was initially unaware or unsure of his eligibility.

■ PRIZES AND AWARDS

The Government, of course, is not the only source of 'free money' for businesses, and it is well for the entrepreneur to be aware of various national and local prizes and awards. At the local or regional level, banks, local newspapers, regional television companies, chambers of commerce and local business clubs frequently offer prizes for entrepreneurism. These are often won by technology-based businesses, and are worth competing for. The amounts on offer can sometimes be quite substantial, and the odd few thousand pounds can be of great value in cash flow terms.

In addition to the cash itself, there is the decided prestige which accrues from winning, which can be of practical assistance if it becomes necessary to renegotiate the major finance for the business. For instance, a firm which won the prize from the local Business Club were convinced that this had a positive influence on their local bank manager's attitude to their financial restructuring proposals.

There are a number of national and regional awards, and the local Enterprise Agency or Innovation Centre should have details. Of particular note are the NatWest/BP Awards for Technology, which are made every eight weeks, and offer up to £10,000 for R&D, prototype or licensing costs. BP also offer Innovation Awards related to the Graduate Enterprise Programme.

There is even more prestige, and prize money, attached to certain national awards, such as the Prince of Wales' Award for Innovation, or the Government's own SMART Awards , which are dealt with on p. 120. These are excellent concepts, and can provide entrepreneurs with a substantial cash injection at the early, developmental stages of their projects.

From the viewpoint of the entrepreneur, which is who this book is written for, these schemes have everything to commend them. However, their ultimate results are at best not proven. A disappointingly small number of prizewinners have developed into viable, growing businesses. The principal reason for this seems to be a concentration on technological 'fizz' rather than the potential market and its penetration. Innovation must *be market-led, rather than technology-led, and it is believed that the managers of such schemes are now actively addressing this question.*

■ GRANTS

Whether or not you are eligible for a grant may well depend on where your business is located.

Many regions of the UK have suffered disproportionately from the economic difficulties of the past decades, and the encouragement of manufacturing industry in these regions has been an important aspect of Government policy. This is partly achieved through special incentives and initiatives for the establishment and growth of smaller firms.

There are currently three types of region: Unassisted Areas, Intermediate Areas and Development Areas. These, and general guidance on the grant schemes available, will now be considered in outline. The major initiatives and schemes will then be dealt with in more detail. Sometimes the rules and the designation of particular regions are changed, so it is as well to check the latest information before assuming eligibility. A map of the current Assisted Areas is shown in Fig. 5.1.

Most people start their businesses from their homes or from the region in which they live. However, the additional incentives available in Intermediate or Development Areas are such that it may be worth considering establishing your business in one of these. This could be especially relevant to entrepreneurs living close to the border of such an area: within commuting distance.

Rather than plough through a mass of brochures and small print, start by contacting your local Small Firms Advice Service, Enterprise Agency, Innovation Centre or Science Park. They normally have staff who know what is available, whether a current scheme is being strongly promoted or whether funds are exhausted, and will guide you through the application. Some organizations make a charge for this, but it can be well worth it, since they can optimize your chances and sometimes find you money which you didn't know existed.

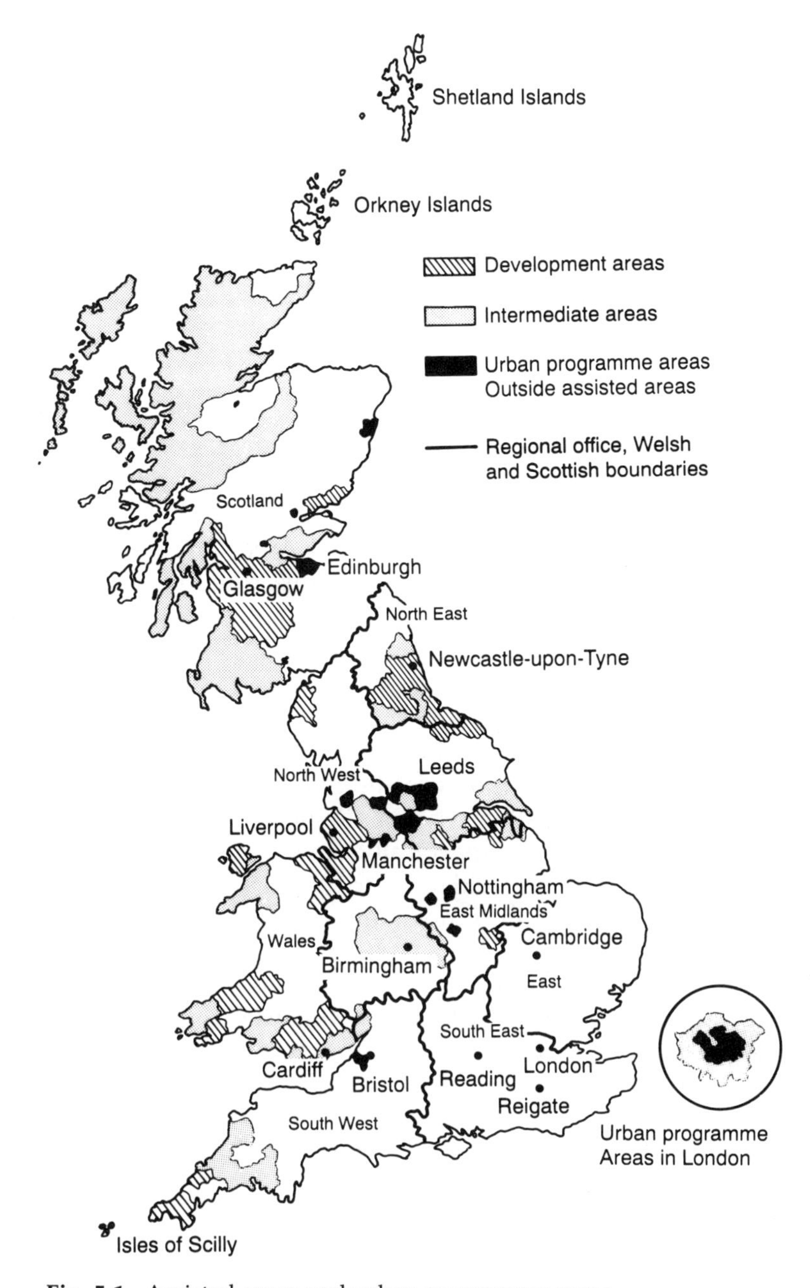

Fig. 5.1 Assisted areas and urban programme areas

UK regions

Unassisted

These include much of the South and South East, together with the more totally rural areas.

There are relatively few grant opportunities available. The DTI Research and Technology support programmes are on offer in Unassisted Areas, and they usually involve collaboration with a university or research association (*see* p. 121). They embrace the various EC schemes such as the main industrial Framework Programmes.

The six Enterprise Consultancy Schemes (p. 122) are also available to firms or groups with fewer than 500 employees. The DTI pays 50% of the costs in Unassisted Areas. The EMRS Export Marketing Research Scheme (p. 124) is also offered in Unassisted Areas.

Intermediate areas

All the schemes on offer in the Unassisted Areas are also available in Intermediate Areas, plus the additional important one of Regional Selective Assistance (RSA: p. 124).

Development areas

These offer the most assistance to the entrepreneur. All the above schemes are on offer, and, for the Enterprise Consultancy Scheme (p. 122) the DTI pays two-thirds of the cost of the consultancy.

In addition to these, there is the Regional Enterprise Grant, or REG (p. 126), which is specifically aimed at new start-ups or diversifications for firms with fewer than 50 employees. The REG includes support for innovation and investment for new developments.

The special problems of Northern Ireland has led to the development of special incentives, which are financially very attractive. These are listed on p. 128.

Major government grant and award schemes

SMART awards

The SMART Awards (Small Firms Merit Award for Research and Technology) deserve a special mention. SMART is a DTI Scheme and the Awards are available nationally. The scheme is run as an annual competition, and firms with fewer than 50 employees are eligible. So are individuals who wish to start a small business. The success of the SMART scheme has led to an increasing number of awards being made available each year: currently, 180.

The Awards are given to support novel, technology-based ideas with a good prospect of commercial success. The award can provide up to 75% (a very high proportion) of first year project costs up to a maximum of £45,000. The winners can go on to win again, which could mean an Award of up to 50% of project costs up to a maximum of £60,000.

This is potentially very useful to the innovator, but the annual nature of the Awards can cause difficulties. I know of one entrepreneur who had his bright idea at the wrong time of the year, and had to wait over six months to hear the result of his application. In the competitive world of technology-based enterprise, this is a long time.

Research and Technology Support

The DTI Research and Technology Support programmes usually involve a research and development collaboration with a centre of excellence such as a university or research organization. They include the Advanced Technology Programme of collaborative research, the LINK programme of research collaboration between industry and research centres in priority areas of research and various European Community programmes of longer-term, pre-competitive research in major areas of technology.

The EC schemes include the main industrial Framework Programmes such as ESPRIT, RACE, BRITE-EURAM and BRIDGE. Their aim is to support Research and Development which is international, collaborative and pre-competitive. This means that it is only likely to be of interest to the very small number of small and medium-sized enterprises (SMEs) that have well-developed high technology expertise *and* the financial resources to undertake long-term research; only around 100 UK SMEs have been involved in the Framework Programme since its inception.

The only schemes of proven accessibility to the SME are the SMART Awards mentioned above (p. 120) and the General Industrial Collaborative Projects, whereby smaller joint development programmes are carried out by the Government's Research Associations on behalf of smaller firms.

SPUR

SPUR means Support for Products Under Research. It is intended to mean Support for Products Under Development, but the acronym (which must reign supreme!) was felt to be less evocative. The scheme, introduced in 1991, offers funding to companies with fewer than 500 employees, at a level of 30%, for development of new products. The maximum is a grant of £150,000 on a total project cost of £0.5 million. The first SPUR award was granted to a company on a Science Park – BDS Biologicals on Birmingham Research Park.

The Enterprise Consultancy Initiative

This has been a very significant scheme in targeting professional business expertise to growing companies. It provides support for expert consultancy help to the specific needs of small firms. Most firms are eligible, provided they or their group employ fewer than 500 people. The scheme has attracted some 80,000 applications for assisted consultancy.

Good, professional business consultancy can be of great assistance to a new and developing firm – but it can be very expensive. The Enterprise Consultancy Initiative provides Government-subsidized assistance to deliver 5-15 days consultancy help in the following areas:

- Business Planning;
- Marketing;
- Design;
- Quality Assurance;
- Manufacturing Systems.
- Financial & Information Systems.

The Consultancy Initiative works in two stages. Following an initial request from your firm to either the DTI, the Scottish Development Agency or the Welsh Development Agency, a so-called Enterprise Counsellor is sent to you to undertake a free, short business review. The Enterprise Counsellor is an independent, experienced 'business GP', and his task is to assess your original request for more specialist consultancy assistance.

Such consultancy may only be carried out by consultants 'listed' by the DTI, and it is acceptable for you to suggest to the Enterprise Counsellor which consultant organization your firm would prefer. If you have no preference, one will be suggested to you.

The Enterprise Counsellor's Report is sent to your company, and also to one of the Scheme Contractors which are responsible for identifying and monitoring the six consultancy subject areas. They normally (but sometimes may not) accept your request for a specific consultant organization. The Terms of Reference for the programme will be agreed between you and the consultants, and the consultancy can then commence.

When the agreed programme has been completed according to the Terms of Reference, you have to pay the consultants between a third and a half of the cost, depending on where your firm is located (*see* p. 120). The DTI pays the rest.

The Consultancy Initiative has been described in some detail, partly because it seems rather complex at first sight and also because it has proved its value in providing independent, high-quality support to companies which are prepared to recognize their needs and try to remedy them. Well over 20,000 projects have been completed since the Scheme's inception. Most client companies had never used consultancy help before, and the degree of satisfaction with the Scheme is high.

The initial reaction to the Scheme was that it would be a 'Consultant's Charter'. Whilst many consultant organizations have certainly benefited financially from the programme, the potential for misuse has been kept at a minimum by rigorous control by the Scheme Contractors. Any attempted deviance from the rules can be punished by 'delisting' the consultant.

The only criticism of the Scheme in practice is that the various stages sometimes make it rather bureaucratic. The initial visit from the Enterprise Counsellor is prompt – normally within 3-4 days of the request – but the time taken to identify the consultant organization and agree Terms of Reference can be inordinate in some instances.

On the whole, however, the reaction 'on the ground' to what is quite an innovative concept has been good. The Quality Assurance Consultancy has been particularly successful, and has led many firms to certification under BS 5750.

SMETAS

This is roughly a 'technology parallel' of the Enterprise Consultancy Scheme, and was introduced at a pilot programme level in 1991. Although it, like the Consultancy Initiative just described, is not a direct source of finance to your business, it does provide subsidized help to sort out essential problems which you otherwise would have to pay for.

SMETAS provides support for a technical advice service to SME, and is principally but not solely focused on the Regional Technology Centres. Indeed, it seems likely that the scheme was introduced to try to enliven these centres, which have not been a dramatic success by any standards.

The Scheme is intended to deal with a range of enquiries about technology and sources of technical information. A call to SMETAS will elicit a visit from a 'diagnostician' (I prefer the term 'technology GP', which is what he or she really is), who will try to solve the problem during the visit. If this is not possible, a technology consultancy can be arranged under similar terms to those of the Enterprise Consultancy Initiative. Whether this scheme will become as successful as the Enterprise Consultancy Initiative is doubtful, but it is still in the early days of its existence at present.

Export Marketing Research Scheme (EMRS)

EMRS is run by the British Association of Chambers of Commerce on behalf of the DTI, and is open to any firm researching the overseas prospects for its products or services. It offers free professional advice on how to undertake export marketing research, and financial support for:

- the cost of commissioning professional consultants to undertake research. EMRS will pay up to half the cost.
- up to half the travel and interpreter's costs of an 'in-house' study, although for some reason this support does not extend to research in European Community countries.
- up to a third of the cost of purchasing certain published market research.

A maximum of two studies may be supported in any year, and, with the exception of the US, only one study per country is allowed. There is a maximum support level of £20,000 per study and a ceiling of £40,000 per firm per year.

It is worth noting that the DTI supports a range of schemes and services for those wishing to export. Some are free, but there is a charge for many of them. The network of Export Development Advisers is especially useful, and new exporters are advised to contact them via the local DTI office.

Regional Selective Assistance (RSA)

The word 'selective' is the operative one here. In principle, there are three types of assistance under this scheme, viz, project grants, training grants and an exchange risk guarantee. The latter is rather specialized, and offers cover against exchange variations on foreign currency ECSC loans (*see* p. 126).

The Project Grant is the main type of assistance. It is related to capital, turnover and jobs. No amounts or percentages are stipulated, other than that they will be based on the minimum necessary for the project to proceed on the basis proposed. The criteria are stated to be:

Viability. The project should have a 'good chance of paying its way'.

Need. There must be evidence that *either* the project would not go ahead if the grant is not given, *or* that it would proceed only on a smaller scale.

Regional or national benefit. Service sector projects which only have a local market are thus out.

Employment. The project should either create new jobs or safeguard existing ones.

Private sector finance. Most of the project finance must come from private sector sources. This includes the entrepreneur's own funds.

One important rider is that *work on the project should not already have started when application is made,* since it is then well nigh impossible to convince the DTI that further support is needed. No commitment should be made until an offer has been received.

Training Grants can support up to 80% of the costs of training young people under the age of 25 in new technology skills: 40% comes from RSA and a matching 40% from the European Social Fund. Once again, the training must be essential to the project. Normal ongoing training and several other types of training are ineligible for assistance.

It is probably true to say that RSA is for diversifications of existing businesses rather than for start-ups. There can be many 'hoops to jump through' and you should be prepared for some hassle and delay. It really is advisable to seek advice before submitting an application. If you don't ask for enough the response can be that you don't really need the grant anyway, but if you ask for too much there is a danger that the project will be deemed inherently non-viable.

The Regional Enterprise Grant (REG)

This, in contrast to RSA, is for start-ups. The Investment and Innovation Grants of the REG Scheme are only on offer in Development Areas, and apply to firms employing fewer than 25 full-time equivalent employees, either in the firm or in the total group of which it is a member. The DTI are quite strict on this: firms with common equity or common directors are deemed to be part of a group. In the view of many advisers, this scheme could usefully have been extended to firms with a larger workforce, say, up to 50 employees; this may now have happened, so check the latest details with your local DTI Office or advisers.

Most manufacturing and most service industries are eligible, which thus includes, say, the consultant or development company in specialised areas of technology. Activities which service only a local market do not qualify. As with RSA, *do not commence your project until you have received an offer of assistance.*

The Investment Grant allows 15% of all capital costs on fixed assets, up to a total grant of £15,000. There is no limit to the project size which may be considered, and eligible costs include plant and machinery (new or second hand), buildings and building costs, and land purchase and site preparation. Capital items on Hire Purchase, where the title passes to the firm, are eligible, but not leased items. Revenue costs are not eligible.

The Innovation Grant is for projects which are innovative to the applicant and which involve a degree of novelty and technical risk. The grant provides up to 50% of the eligible costs up to a maximum of £25,000. The eligible costs are currently broadly interpreted, and include all development costs including capital costs, up to the point of commercial production. Some costs of marketing – currently, up to 15% of the grant paid – are grant-eligible.

These grants are straightforward in practice, involve a minimum of 'red tape' and have been highly appreciated by the recipients. The concept of the Innovation Grant is particularly appropriate, since it recognizes that, in a technology-based innovation, there is a 'technical risk' factor over and above the normal commercial risk of new projects. The Grant is an attempt to share this technical risk.

Cheap money

A number of organizations offer loans under very advantageous terms in order to support business development. Most of these are regionally constrained, for instance to the regions affected by closures of steelworks and coal mines. If you happen to be living in such an area, these loans and other special schemes are worth pursuing. Any Enterprise Agency or Innovation Centre in the region will have more information.

■ ECSC LOANS

European Coal and Steel Community Loans provide fixed interest, medium-term loans for projects which create new employment in areas afflicted by closures in the coal and steel industries. These can be for up to 50% of the fixed assets, are normally for eight years and there can be a four-year moratorium on the repayment of capital. Any manufacturing or service industry can apply, and some successful technology-based firms have been assisted by such loans. These grants can take a long time to be agreed, and applicants should apply in good time, well before the start of the project.

■ BRITISH STEEL (INDUSTRY) LTD

British Steel (Industry) Ltd. (BSCI) provides business finance in the 19 traditional steel areas, with the aim of diversifying and strengthening the local economies. Both loan finance and share capital is available, typically in the range £20,000 to £70,000.

Help is principally directed to existing manufacturing businesses, and the owners are expected to provide a significant financial contribution together with support from a bank or other source of private sector finance. However, it can be beneficial to seek help at an early stage, since, if BSCI like the project, their influence on other financiers to make funds available can be quite considerable.

Loans are of a typical nature, at a fixed rate of interest close to bank base rates and for a 2-4 year period, but there is the possibility of a capital repayment holiday.

Equity capital is normally on the basis of preference shares with an exit after about five years.

A **Seed Capital Development Fund** is also offered as a means of helping companies to commercialize new products, and this is particularly relevant to start-ups. It normally takes the form of an unsecured loan of up to £25,000 at an interest rate well below commercial rates. The repayment period can be as long as five years, and it is possible to negotiate a capital repayment holiday of up to three years.

■ BRITISH COAL ENTERPRISE

British Coal Enterprise (BCE) has a similar loan scheme to British Steel (Industry), and it has also introduced an equity scheme, providing funds between £20,000 to £100,000. Their objective is to find the 'missing money' that will make a project go. In other words, if you have a viable project and have acquired a substantial proportion of the finance needed from yourself and other private sector sources, BCE is prepared to find the rest. Like British Steel (Industry), they can also take the lead in finding finance for a project, which can be to the entrepreneur's advantage in securing other financial backers.

BCE works in the traditional coal mining regions, with the same objectives as British Steel (Industry).

■ LOCAL AUTHORITIES AND DEVELOPMENT AGENCIES

If you enquire at your local Town Hall or Local Authority Industrial Development Unit you may find that they have 'seedcorn' or low-interest loan schemes aimed at encouraging enterprise within their region. So do Scottish Enterprise, the Welsh Development Agency and the Local Enterprise Development Unit in Northern Ireland. Since these three agencies were formed to encourage economic and industrial development in their respective regions, it is worth briefly summarizing the financial schemes for technology firms available through them.

Scottish Enterprise

The SDA have produced a booklet with comprehensive descriptions of schemes available. The Agency itself offers loan and equity capital, but with the objective of encouraging maximum private sector support. The scheme favours areas of high unemployment and rural areas suffering from depopulation.

The Royal Bank of Scotland has a technology fund which is specific for firms starting or expanding in the city of Dundee. This takes the form of a low-interest loan between £15,000 to £100,000. In the first two years, interest is charged at only 5% per annum; it then gradually rises over the period of the loan. It is specific for technology-based companies of the kind described in this book.

The Welsh Development Agency

The WDA have extensive services to support business development. Their Technology Growth Fund offers support for market-led innovation to firms in Wales, with support tied to the development of a specific product rather than the whole company.

Local Enterprise Development Unit (Northern Ireland)

LEDU in Northern Ireland offer some attractive financial packages to encourage growth of small firms. Capital Grants of up to 50%, Interest Relief Grants to reduce the cost of money borrowed from the private sector, Rent Relief Grants and Employment Grants are all available. There are also Setting-up Grants which can contribute towards planning fees, legal and accountants' fees and removal expenses. The amounts for all these are related to the number of jobs created or safeguarded.

Innovative businesses attract further incentives. Research and Development Grants of up to 60% are offered to selected projects. There are in addition Market Development Grants and Management Salary Grants (to strengthen the management of SME by employing specialist managers). Secured loans at reduced interest rates and with interest and capital repayment holidays are available, as is equity capital with an agreed buy-back option.

Chapter 6
SOURCES OF FINANCE: LOANS AND EQUITY FINANCE

Bank loans

This Chapter continues the subject of Chapter 5, moving onto cover the more 'expensive money' of bank loans and equity finance.

Bank loans were rated in the list in Chapter 5 p. 115 as 'Expensive Money'. The banks and financial institutions cannot live on 'special offers' any more than the local supermarket. Since bank loans and equity financing constitutes the largest source of finance for growing businesses, you must expect the bulk of your financial requirements for developing a business to be at full price.

However, it is still important to shop around. High street shops do not all have identical products or prices, and neither do the banks and financial institutions. Some 'package' their offers in forms especially suitable for particular classes of entrepreneur, since the financial needs of someone starting a printing shop are different from someone developing applications for monoclonal antibodies. The major schemes relevant to technology-based firms are considered on p. 133, and a list of contact points are appended.

It seems likely that as we settle into the nineties, the lending policy of most banks is likely to be more restrictive because of the number of past bad debts. Closer scrutiny of applicants for investment loans will be an inevitable result. The banks have also received criticism over their treatment of small business customers and have smartly brought out codes of conduct and 'customer charters'; this probably means that if you *do* get a loan, they will treat you fairly, to avoid further opprobrium.

A bank has the virtue of being accessible – there is one in every high street – and potentially flexible. In principle, they can offer everything from a substantial long-term loan, involving a repayment period of up to 20 years, to a short-term temporary overdraft. The interest can be at a fixed or variable rate, possibly dependent on a number of factors. But they will normally look for some kind of security. Let us now look at these three factors in turn.

■ TERMS AND CONDITIONS – THE 'SMALL PRINT'

Repayment terms

There are a variety of repayment methods. They range from arrangements to repay at specific times agreed by the bank, to regular instalments, to repayment when the period of the loan expires. Many banks now have special schemes to help new and smaller firms, and it is worth asking about these, even if they are not overtly offered to you. They include deferred repayment of capital, lower interest rates for the first year, or complete repayment 'holidays' over the first few months or so.

Such schemes can be useful to allow a business to get 'off the ground', but it is wise not to rely entirely on this kind of financial juggling. It is better for your Business Plan to be sound without such adjustments, and then to consider a deferred payment scheme to make the first year just that much easier. Don't forget: the banks will want their money back at some stage, and you are likely to be eventually paying more for the privilege of being allowed to deviate from the norm.

If the business is unexpectedly successful in the earlier stages and generates a lot of cash, you will want to be able to pay the bank back at least some of the loan early: check that there is no clause either forbidding this or invoking a substantial penalty.

If the business fulfils all the aspirations of the Business Plan, then your need for finance is likely to increase rather than decrease, in order to finance the next stage of your growing company. There are then other options open. You might simply be able to pay back your original loan and substitute a larger one. Or you may be able to convert the outstanding loan into shares, with the lender thereby taking some of the equity. At this stage you need expert financial advice.

Interest rates

There are two principal options here. The first is the fixed rate, which is constant over time. It may not always be offered, but if it is, it insulates you from the vagaries of market fluctuations. This is a mixed blessing, since it can prove very expensive if the market rate falls and stays low for any period of time.

The second option, of an interest rate which fluctuates at a specific level above the bank's base lending rate, is more common. This offers an easier time if base rates are low, but can produce problems when they rise and stay high for any period. During 1990, when UK industrial growth was very low, interest rates, which had risen throughout 1989, were first kept high to induce lower inflation. They were then maintained at a high level when Britain joined the European Exchange Rate Mechanism. Whatever the merits of this course of action – and the top experts seem far from agreement – it caused real pain amongst small and medium-sized firms, and many failed.

Do not be misled by the bank manager who tries to give the impression that an interest rate is a single, immutable figure, sent down from on high. Do not be apprehensive about trying to shop around, to find a bank which is more sympathetic to your particular project. An interest rate reduction of a half to one percent can add up to a lot over the period of the loan.

Equally, be wary of those offering finance at significantly lower rates than the major banks. There are probably hidden costs involved which will easily outstrip the savings made by virtue of the lower interest rate. Such organizations will almost certainly fail to understand the nature of your business, and you will be laying up trouble in store for the future.

Within the overall fixed or variable rate options, a number of variations may be possible. It is possible, but uncommon, to have the interest rate linked to some measure of your company's results, such as sales or profit. More commonly, banks sometimes have special schemes associated with types of companies or with centres of business growth such as Science Parks and Innovation Centres. It is worth finding out about these, since they can save on the interest repayable.

Security

In one sense, a first-rate business plan is the best security a bank can have. It encapsulates within it a marketable product, a sound management team and a realistic financial forecast. However, don't try telling your bank manager this: it is not enough! The bank will normally want more tangible security or collateral, which means either a charge over the assets of the business, or a charge or personal guarantee over your own assets.

Do not expect the bank to lend against the full value of newly acquired assets. It prefers not to invest more in a project than the owners, which brings the maximum borrowing limit down to half the total required.

In addition, technology-based businesses often have few tangible assets with a high resale value, and many entrepreneurs are faced with the choice of either subjecting themselves to substantial personal liability or abandoning their project. At a later stage, the business's debtors will be regarded as an asset, but at the start-up or product development stages these will not exist.

If there is a charge over your business assets and you default on payment due to the bank, the assets will be sold to pay the bank (usually by an auction under 'forced sale' conditions) before the other creditors are paid. It will also be difficult to sell these assets for any other reason if a charge remains upon them.

■ SPECIAL SCHEMES

The Loan Guarantee Scheme

The Government's Loan Guarantee Scheme is worth mentioning. The Scheme was established some ten years ago to encourage the banks to provide medium-term finance up to £100,000 in circumstances where the risks were too great to warrant more conventional loans.

It was aimed at those firms where insufficient security was available, and, under the Scheme, 70% of any such loan is guaranteed repayable to the lender by the Department of Employment if the borrower defaults. It therefore seems particularly apposite to businesses involving a degree of technical risk, where current assets are not sufficient to provide sufficient security for a normal bank loan.

The Loan Guarantee Scheme is operated through the main clearing banks plus a number of smaller banks. It is not intended to compete with normal commercial finance, so it is not available if a conventional loan can be obtained. If the lender decides that a firm is eligible, the lender applies to the Department of Employment for a Guarantee. In return for this, the Department levies a premium of 2.5% per annum on the amount guaranteed from both the borrower and the lender. There is a 'fast track' for loans under £15,000, and if you plan to establish your firm in an Inner City Task Force Area, the Guarantee increases from 70% to 85%.

The Scheme has undergone several modifications since its inception, and has never really been a great success. It is expensive to run, is inflexible, and excludes a range of activities (none, however, specifically related to technology-based businesses).

A major drawback stems from its principal aim of providing security. Borrowers under the Scheme find that banks will not lend them further funds, since this questions the need for the Guarantee in the first place. Another is the cost of the premium levied by the Government on the loan. Finally, Loan Guarantee Scheme funds cannot be used to recover money already invested or to pay off current liabilities.

The fact is that security never made a bad proposition good, and there have been a number of expensive failures of the Scheme because this basic principle was ignored. Nevertheless, the Scheme is available, and it has covered lending of over £700 million to more than 21,000 small firms; it is thus worth bearing in mind.

Specific schemes for technology-based firms

Although past experience shows that bank managers who understand a technology-based business are few and far between, some of the major clearing banks have now recognized this deficiency and have initiated a number of special services geared to small firms dealing with high technology products and services.

Barclays Bank have around 60 specialist high technology branches located throughout the country, backed up by specialists in their London-based Corporate Division. They back this expertise with a set of funding schemes whose objective is to provide flexible forms of investment suitable for technology-based small firms.

The Midland Bank has its Regional Business Advisers plus a team with technology experience. Besides the usual loan facilities, they have an arrangement for the purchase of equipment and machinery which entails no arrangement fees or other costs.

The National Westminster Bank, with its Technology Unit, has moved to support technology-based businesses in a substantial way, and it is encouraging to note their willingness to listen to the particular needs of such companies. The Bank has established a network of 'technology literate' managers and advisers throughout the country. They have also, partly through the United Kingdom Science Park Association, developed specific links and schemes with and through Science Parks and Innovation Centres.

In particular, NatWest's venture capital arm, County NatWest Ventures Ltd., offers seed capital in the form of *unsecured* loans to technology-based businesses, in exchange for a shareholding agreement, i.e., the Bank requires an option on a minority shareholding in the company. The scheme includes the possibility of interest or capital repayment holidays of up to four years. A technical appraisal of the project is usually required. This service is called the New Technologies Appraisal Service and is available to help NatWest Branch Managers make a funding decision about a technology with which they may not be familiar. An appraisal is undertaken (if considered necessary) by a number of specialist consultants, and the cost is usually £1,800 of which NatWest will pay half.

This is an interesting new approach, which is designed to assist branch managers make decisions on a technology which they may not be familiar with – a problem known to many who have started a technology-based company.

Lloyds Bank does not have any financial schemes specifically designed for the technology-based entrepreneur, but it does have an Innovation and Technology Advisory Service (ITAS), established in liaison with PERA, the Production Engineering Research Association. ITAS is a consultancy service for which fees are charged, for market research, information technology, CAD/CAM, materials engineering and advanced manufacturing technology. Before using this, the entrepreneur might well be advised to consider the DTI Consultancy Initiative, SMETAS or SPUR (*see* p. 122) as possible alternatives.

With such a plethora of new schemes, some well-meaning but impractical and some still in their infancy, it is difficult to give definitive advice. The best suggestion is to find out what is on offer, soliciting advice from the local Small Firms Service, Innovation Centre or Science Park; and do not expect your local bank manager, possibly one whom you have dealt with for years, to necessarily give you the best deal available.

■ WHAT THEY WANT FROM YOU

Having reviewed what is currently on offer, let us now look at the matter from the bank's viewpoint. Most bank lending implies strictly limited rewards for the bank; there is no additional pot of gold for it, even if the borrower is fantastically successful. A bank manager responsible for arranging business loans therefore wants to be sure that the bank will get the money it lends back plus the interest, and that it will get it at the time it is due. Since the bank is probably not in it for anything above this (it is not normally an equity-holder and will not have a share in the overall profits), banks tend to be conservative in their assessment of businesses. They focus on the risks rather than the opportunities.

In order to minimize risk and maintain security, the bank will look at four areas. The first is soundness of the Business Plan. We have dealt with all this elsewhere, but it is worthwhile to emphasize again the value of the Business Plan, and the *paramount* importance placed upon it. The principal areas of investigation will be:

- the track record of the management team, particularly that of the principal entrepreneur or project 'champion';
- the technical performance parameters of the product, in particular

the innovative characteristics which might give a marketing advantage;
- market information: its size, growth or potential growth, how the product or service will be promoted and sold and how the competition might react;
- the skills available, with particular consideration of whether the complete range of complementary skills are present;
- overall timescale.

The other areas the bank will investigate are more directly financial, and are dealt with below.

Gearing

Gearing is the ratio of the borrowed funds in a company to shareholders' funds; the latter includes both the initial share capital and any subsequent profits which remain unclaimed. A firm which has a lot of money invested from external sources compared with that of the shareholders is said to be highly geared.

The banks prefer not to have more money in the business than the shareholders themselves.

Gearing is thus an assessment of the owner's relative financial stake in the firm. It is perhaps not surprising that banks generally like to see a commitment from you that is comparable to the one you are seeking from them. This is another problem for the technology-based project, since the entrepreneur may well have the technical and managerial ability but simply not have the cash available to satisfy the bank.

It would do little harm if the banks occasionally took a more flexible attitude towards this, and there are signs that things are now changing. If the entrepreneur, project and Business Plan are sound, there seems little point in the banks insisting that the entrepreneur puts in a significantly larger proportion of the cash required. If he has already invested all he can sensibly afford, all this additional commitment does is to increase his personal anxiety and consequently reduce his ability to manage at the highly critical start-up time. The bank manager should exercise both judgement and tolerance on such occasions, and allow the gearing ratio to rise above the 1 : 1 level.

An alternative is to persuade an additional shareholder to come on board, and the entrepreneur should not be averse to this, if it enables the project to go ahead (*see* p. 138).

There is another effect of gearing which increases the risk to the bank. The interest payments you will have to pay do not depend on your company's performance. They are dictated by interest rates which in turn are determined by macro-economic factors. So, the same amount must be paid, even when the company is going through a bad patch; and the more highly geared you are, the higher the risk to the bank.

Matching of the loan to its purpose

The banks have a variety of loan schemes for different purposes, and you should recognize this in planning your approach to them. As a general rule, the sum lent should be commensurate with the cash flow expected from its application.

For instance, the bank would expect a medium or long-term loan to be used for capital or fixed assets, and not for working capital: the latter is better accommodated by a short-term loan or a temporary overdraft facility.

Security

The requirement for collateral has been dealt with earlier (*see* p. 131); it stems from the bank's need to get its depositors' money back. The bank will thus attempt to have security throughout the whole term of the loan. *Do not enter into personal guarantees without the most careful advice and thought, and without discussing it with your family.* Any charge over your personal assets can cause untold hardship – and not just financial ones.

I have seen promising entrepreneurs suffer enormously from risking too much at an early stage. Inevitably, things did not go as well as they hoped, and this brought financial difficulties. The financial difficulties can then easily lead to family problems: the family of a comfortably-off manager, who tries to start out on his own, do not always take kindly to leaving their house which is pledged to the bank and moving to much less attractive rented accommodation.

Unfortunately, it does not stop there. A technology-based business depends critically on the managerial and intellectual skills of the entrepreneur. If he or she is beset with financial and personal difficulties, they can rapidly become all consuming to the total detriment of the business and quite often the marriage and health of the entrepreneur as well. The guaranteeing of personal assets in support of weak business proposals has been strongly criticized in an investigation of failures from the Loan Guarantee Scheme.

So, be ultra careful when faced with the possibility of pledging personal assets. Perhaps the best advice is don't do it unless you are sure you are prepared to lose it and can afford to lose it.

The entrepreneur who has read thus far may be tempted to give up and go and grow dahlias instead! However, remember that many technology businesses do succeed and become highly profitable. It is quite normal for bank lending to go smoothly, and, when it does not, the banks can be enormously helpful. A lot depends on developing a positive relationship with the bank manager. This means that you must try to appreciate what the bank is looking for, and the bank must try to understand your business and its true potential.

■ WHAT YOU MUST BE PREPARED FOR

Forewarned is forearmed, and it is important to give some forethought before you enter the bank to seek finance. It is assumed that your Business Plan is a good one, and will have already convinced the manager that you have thought long and hard about your product, its market, your management team and your cash needs.

Be prepared to negotiate

'Be prepared to haggle' might be a better heading. Don't overdo it, but it is worth a bit of hard bargaining to get the kind of deal you can be comfortable with. As indicated earlier, negotiate over the interest rate and collateral required. The bank will be looking for security, so that whatever happens they will get their money back. You should avoid giving up too much collateral, since it puts you in a restrictive position in which your ability to borrow further in the future will be curtailed.

If you think that the bank or its manager is totally failing to understand the nature and opportunities of your business, then look around and find one more attuned to your needs.

Check the incidentals

Arranging a loan costs money. Besides any fees which you have to pay to your accountant or financial adviser, the bank will often charge fees for arranging the loan, which go under various guises such as arrangement fees and management charges. You should foresee these, but find out what they are likely to be before you come face to face with the unexpected size of them.

Equity finance

The total share or 'equity' in a company represents the amount of money which has either been given or promised by its shareholders. A share is a unit of such capital, normally with a nominal or face value of £1.

In simple terms, the shareholders acquire certain rights in exchange for investing their money, which may include:

- **a dividend**, i.e. a proportion of the firm's distributed profits in the ratio of the number of shares held to the total number distributed;
- **voting rights**, which gives the shareholder some say in the overall business;
- **some rights to the assets of the company**, should it go into liquidation.

There are three principal sources of equity finance: the individual investor, the institutions (venture capital) and other companies. The involvement of other businesses, in what is known as joint venturing or corporate venturing, is dealt with later on p. 156.

■ WHY SELL OFF ANY OF THE COMPANY?

If you are a technology-based entrepreneur, the question you probably first ask yourself when faced with the idea of raising money by selling shares is: why should I give anyone else a share of the action? If I had the idea in the first place, have taken all the risks so far, worked all hours and suffered all the blood, sweat and tears, why should I give up any of the ultimate rewards? It was *my* idea, *my* risk and it should be *my* profit.

This attitude is understandable, and one can have some sympathy with businessmen who take this stance. They have perhaps come out of a relatively secure job, have risked their family's welfare, their money and often their physical and mental health to get a business under way. Why should they be forced to give up a substantial part of what they fought for, just at the time when it is going well and is ripe for expansion?

A growing company is different from a start-up

Some business advisers lay stress on the difficulties of start-up, and, indeed, these are not inconsiderable. In my view, however, the subsequent stage of growth can be at least as traumatic, if not more so. The change of 'culture' and managerial attitudes is discussed in Chapter 1, p. 9, but one of the prime changes stems from the need for investment – sometimes lots of it – and consequent selling of equity.

One of the first difficulties to be overcome is the mental one of the entrepreneur coming to terms with the fact that parting with total ownership is frequently the *only* option. For technology-based firms the choice is often stark. Firstly, there is the need for substantial investment or the firm cannot grow. The option of staying small will not work, because the product is high-tech with high added value, and the competition will soon be along to erode the market share. So it is either invest, or perish.

Secondly, the firm has probably already borrowed quite a bit from the banks, and they will not lend any more. Despite your impeccable record in paying back the loan, and the fact that you have tenaciously kept to your Business Plan so far, the banks, with their thoughts principally on getting their money back (*see* p. 134) will be reluctant to allow the gearing to increase further.

This Section is really breaking the news, as gently as possible, that selling off some of your company may be the only way ahead. Putting it bluntly, it is better to accept 70% (say) of a viable, expanding business than to be left with 100% of a failure.

This does not mean, however, that you should rush into the arms of the first investor that shows an interest and accept any deal that is offered. Investors will surely weigh you up, and you should also assess their offers and make sure that they not only offer you money, but offer it at the right time and on the right terms. These aspects will be considered below.

There are also some considerable benefits in seeking equity investment. This is not just stated to 'sugar the pill' of giving up some of your company. There are positive aspects to having equity investors, over and above the injection of cash they bring with them.

Equity investment is higher risk money

The banks want their money back, with the agreed interest and at the agreed time, however the business happens to be faring. Their money is therefore relatively low risk. An equity investor, however, is in it for better or worse, shares in business failures as well as successes and does not normally get anything back if the business fails.

Equity investment is higher growth money

Because of the high risk involved, equity investors seek higher growth rates rather than income from distributed profits. The fact that no dividends are required means that the cash can be used in the business instead. This 'breathing space' or 'buffer' may be highly desirable to a firm which needs some time for product development, and cannot easily afford the heavy initial repayments which loan financing would entail.

Equity investment maintains borrowing ability

In general terms, it is good to maintain the business in a state where you can still borrow more if you need to. If you borrow up to the limit, you will not be able to respond to a sudden need or opportunity; you will also, in this condition, not be sought out by potential equity investors.

In other words, if you raise finance through equity, the gearing of the company is kept lower, and your ability to borrow is not adversely affected.

Equity investors can be good advisers

Experienced investors have a good knowledge of financial matters and could be useful to a small and growing company, both as a direct source of advice and as a link to other expertise. They will want to get the best from their investment, which will consequently benefit your own.

Some investing organizations take a 'hands-off' approach, but many will insist on an active role, probably through the appointment of a non-executive director, in the direction of the firm. If carefully considered by both parties, this can be of considerable benefit to the company. The pros and cons of this are considered later (*see* p. 150).

Equity investment is more easily adaptable

Loan financing is slowly becoming more flexible, as the banks realize that the schemes they had were not generally suitable to the innovative small business, but this form of financing is inherently constrained by the risks involved and the banks' attitude towards them.

Equity finance is not so inhibited: there are more parameters which can be varied. In theory, therefore, it should be easier to structure the deal to suit the specific need. In practice, there is still a long way to go, especially when it comes to financing the smaller firm with a need to support a development programme before it can generate any substantial sales income.

Nevertheless, there is still more room for manoeuvre in equity investment, and it is possible to negotiate over the rights of shareholders to accommodate the needs in specific cases.

Sources of equity finance

■ PRIVATE AND SMALLER INVESTORS

Family and friends

Maybe you have someone in your family or amongst your friends who has sufficient capital, and wishes to invest in your project.

If so, it can make good sense to invite them to participate, but make sure they go in with their eyes open. Be certain yourself that you want them in the project. The opportunities for family feuds and quarrels in the boardroom are legion! Perhaps the best suggestion is that you *insist* that they take independent expert advice before investing. The risk, as with any lender, should match the potential reward.

Others

There are always people looking for a good investment, and some of them are not only looking for a return on their capital, but also have an altruistic intent, such as a desire to help new businesses in their own region. Such philanthropists are usually known to local Training and Enterprise Councils, Enterprise Agencies, Innovation Centres or Science Parks, and contact with such centres is the best way to find such investors.

There are several legal problems related to linking small firms with small investors, but a number of 'marriage brokers' do exist which are able to provide such a service. These again will be known to the local enterprise network. The Government has asked the Training and Enterprise Councils to set up a network of contact points, in a two-year pilot scheme involving five areas of the UK.

One network which is worth a special mention is the Local Investment Networking Company, or LINC, which was formed in 1987 by a number of local Enterprise Agencies to provide a national scheme. It now comprises 15 agencies, and for a fee of £50 it will attempt to match entrepreneurs seeking amounts up to £150,000 to potential investors. Although a glance at their regular Bulletin reveals that relatively few of the projects are substantially technology-based, there is now a scheme run in conjunction with BP called the BP Innovation LINC, which offers advice and investment between £5,000 and £50,000 to encourage early-stage technical businesses.

Indications are that investments *are* made through such intermediaries, but that technology-based firms are not the prime beneficiaries. Although the amounts invested are relatively small, which is the most frequently needed level of financing by such enterprises, the lack of understanding of technology by the investor tends to be a drawback.

The Business Expansion Scheme (BES)

The BES was established in 1984 as an encouragement to individuals to invest, on a reasonably long-term basis, in unquoted UK companies. The investor can claim full relief on income tax on investments kept for five years or more.

The history of BES is by no means unchequered, and there have been a substantial number of investors who have lost their money, either directly or through an approved BES investment fund. The latter are established to spread the risk over a portfolio of investments, and are professionally managed; they are thus no different from other venture capital funds from the entrepreneur's viewpoint.

There is no need to review the Scheme from the investor's angle in this book. The point is that BES investors are people paying income tax at the higher rate, and would prefer to pay Capital Gains Tax, currently at a lower rate. They thus also wish to see the value of their investment grow over some years, rather than receive several dividends.

This could be very appropriate to your needs as an entrepreneur, since your Business Plan may well indicate that all the money generated needs to be put straight back into the business, rather than in paying dividends. The fact that most BES investors are interested in smaller sums than the major venture capital funds also makes the Scheme relevant.

There are other possible advantages. BES money can often be raised quickly. Since the BES investor has to hold his shares for at least five years to get the tax relief inherent in the scheme, the company enjoys a stable shareholding. BES investment can widen the shareholding base of a firm without the necessity of the existing shareholders relinquishing financial control.

There are, however, disadvantages. The major one is the tendency for BES funds to be invested in 'safer' things like property, retailing or conventional light industry: areas the small investor understands. There is a noticeable paucity of investments in start-ups or technology, which is really what the Scheme was intended for.

Further problems include:

- difficulties in obtaining further finance, since the original BES-shareholders may lose their tax-exempt status if they invest more.

The company must also satisfy BES legislation for some years, which may jeopardize its chances of securing additional finance from more conventional sources

- charges and fees to BES Fund managements
- the detailed legislation involved, which is quite complex
- restrictions to trading wholly or mainly in the UK. Anyone planning to expand overseas should probably seek investment elsewhere. This could well affect many technology-based innovations, which often need an overseas involvement to produce a suitable size of market for viability.
- possible requirements for equity options. Some BES funds want an option over 10-20% of the equity, implying that, if the company is successful, the ownership of your business will be lessened.

Such difficulties are reflected in the figures, which show that funds encouraged by Government action such as the BES are contributing less and less to the total amount invested by the venture capital industry. The failure of 24,000 businesses in 1990 has also been a disincentive to the 35,000 BES investors in the UK. The scheme currently appears to have a limited lifetime.

■ VENTURE CAPITAL

It is easy to forget that the venture capital industry is still relatively new in Great Britain. In 1979 venture capital investments totalled around £20 million. By 1987 the figure had grown to over £1,000 million and today there is over £6,000 million available for investment from such funds. This, as the British Venture Capital Association says, is 'big business for Britain', and the total funds available exceeds the sum of all other such investors in Europe. Much of the money comes to the venture capital funds from the banks, pension funds and insurance funds. Although most of these funds are in 'safe' long-term investments, a small amount is allocated to the riskier but potentially higher-return venture capital funds.

For a variety of reasons, there are real problems facing any technology business seeking venture capital. These problems should be faced, both by the entrepreneur *and* by the venture capitalist. There is evidence that good, technology-based projects are failing to secure the necessary investment because of a lack of entrepreneurism on the part of the UK financial industry. Perhaps the review here will encourage some fresh thinking on the subject, so that funds and managements will emerge which adopt a longer-term view and a more enterprising spirit.

■ PROBLEMS FOR THE TECHNOLOGY-BASED COMPANY

The nature of the problem

Much of the rest of this Chapter is devoted to advice on what the typical venture fund is looking for, and what the entrepreneur needs to do to present his case. However, the facts are that comparatively few small to medium sized technology-based companies have acquired funds through equity investment. The problem needs to be recognized in the short term by the entrepreneur seeking funds, and in the longer term by the venture capital industry, which needs to counteract the 'short-termism' which currently besets the UK.

It would be nice to present a clear statistical analysis of the problem, and some data are cited below. Hard statistics are difficult to come by, however, since firms that are repeatedly spurned by venture capitalists either give up, increase their bank loans to unwise proportions or limp along, never achieving their true potential for themselves or the national economy.

Anyone dealing with the problem of seeking investment for small, technology-based firms knows that the venture capital industry has currently little interest in them. One of them also told me. 'We have assembled a £30 million fund for technology but we have invested only £5 million. The rest is in the bank earning 13%, and it has to be a pretty good investment that can give our shareholders better than that.'

Another said to a small, technology company: 'Your problem is partly in your company's name, which includes the words "new" and "technology". We're not interested in anything that's new and untried, especially if its technical.'

In 1991, I was assisting a soundly-based and growing firm in the medical and health care area to acquire around £0.5 million growth-phase finance. I contacted a venture capital company whose stated interests were in health care and technology, and whose preferred level of investment ran 'from £0.5-£2 million for all stages of company growth.' I was told that they in fact rarely invested less than £0.75 million, and only in firms that were already profitable, with an established market. Technology was currently 'out'.

It came as no surprise, therefore, to find the National Economic Development Office (NEDO) stating in January 1991 that 'it was felt that financing problems had constrained the development of biotechnology in the UK.'

The size of the problem

Some aspects of the problem are illustrated statistically in the results from the 1988 study by Monck, et al. of 284 technology-based firms (Table 6.1). These indicate that the number of companies with equity from venture funds were limited: only 3% for start-ups and 8% for growing companies.

A number of explanations were advanced for this, including the reluctance of entrepreneurs to accept an equity partner. Another reason was the doubt whether the technology-based firms have the required management capabilities. There was also an uncertainty among fund managers as to whether there would be a satisfactory exit route for their investment, so that the investment could be realized and the funds reinvested elsewhere.

Finally, there was a regional effect: fund managers prefer to invest in firms located close to them (which means in London and the South East), in order to maintain regular contact.

Yet there is no shortage of venture capital, in the UK or the European Community as a whole. Figures from the British Venture Capital Association tell us that they invested £1,647 million in 1989 in 1,569 companies. They have, they inform us, supported a significant number of start-ups: some 13% of financings in 1989, with the average size of the start-up financing involved being £486,000. There was also substantial support for 'early stage' financing, at an average level of £375,000.

This level of early stage funding, however, is five to 10 times the amount often needed by a technology-based start-up, and is usually destined for manufacturing industry or large retail outlets (some of these have not proved to be very wise investments in the last few

	Percentage	
Source of Finance	Start-up	Growth
Personal savings	55	20
House mortgage	2	1
Existing business	6	4
Retained profit	–	26
Clearing bank	17	25
Venture capital	3	8
Private equity	2	3
Public agency	4	4
Grant	1	3
Loan Guarantee Scheme	2	3
Other	8	3
TOTAL	100	100

Table 6.1 Sources of finance for technology companies.

years!). The principal early need for innovative technology businesses is for investment in the £100-£300,000 range, and there is a clear reluctance within the venture capital industry to supply such capital.

The BVCA Report notes that over 20% of companies financed in 1989 were technology related, and the amount invested rose significantly, so perhaps there are signs of improvement. Against this is the strong regional bias, with London and the South East accounting for 39% of companies financed and 49% of the amount invested. There is little evidence of any increase in support to the 'regions'.

The reasons for the problem

A number of explanations have been given and several papers written about this mismatch between seed capital needs of the small, knowledge-based firm and the objectives of the typical UK fund and its managers. Undoubtedly, there are a multiplicity of factors, but some are either given undue prominence or are just plain spurious. For instance, there are undoubtedly small firm managers who will resist all attempts to get them to give up a single percent of equity. But there are still a significant number of entrepreneurs who would be ready to do this, and many of these are totally disillusioned about the ability of venture fund managers to begin to 'get on their wavelength'.

It also seems inherently unlikely that entrepreneurs operating 'North of Watford' are any less worthy of support than their counterparts living in the South East. The problem seems to be – with one or two honourable exceptions – the marked reluctance of venture capitalists to venture far out of London, unless it be to another European capital.

Short-term horizons

One reason is the problem of 'short-termism'. We are beset in the UK with the demand for instant success and rapid profits. We have, in the late 80s and early 90s focused attention continually on shorter and shorter time scales. Any Business Plan from a technology-based project which dares to suggest that cumulative profitability will not occur until after two years is regarded as suspect, however sound it may be in the long term.

Harry Nicholls, the Chief Executive of the venture fund associated with Aston Science Park, Birmingham Technology (Venture Capital) Ltd., has made his attitude to short time scales clear: 'All the investments at Birmingham Technology, Ltd. are in high-tech start-ups, and we expect to take seven years on average before we get realizations; allowing for lead times and some failure, this may mean 12 or 13 years before the fund is fully mature and it becomes reasonable to evaluate performance – a very long time by most standards... However, this is a 10-15 year development business, as experience in the US has shown.'

'Short-termism', which leads to venture funds seeking very high rates of growth such as 40% p.a. from the start, gives rise to the feeling often expressed by entrepreneurs that fund managers 'are just not talking the same language'.

Assessment costs

The cost to the fund, in terms of management and time, of assessing potential investments is largely independent of the size of investment required. In fact, the management needs for a smaller firm may even be greater. It is thus inevitable that many venture funds prefer the larger individual financings, and is perhaps an understandable reason for their reluctance to look at many worthy plans from smaller firms.

Management capabilities of small firms

There are doubts amongst the venture funds as to whether the smaller, technology-based firms have the management capability to achieve the levels of growth and profitability sought by the fund. This doubt is fed by a variety of factors, some real and some imagined.

The management capability may be genuinely lacking in quite a number of instances. Although management skills in smaller firms have improved over the past five years there is still a long way to go. Of all the various reasons – I almost said 'excuses' – advanced by venture capitalists for failure to invest, this is the one which rings true. The acquisition of management skills is a vital matter for all small firms, and the current emphasis on training is absolutely imperative.

A way ahead

Some within venture capital assert that small firms should build strong management teams, and prepare themselves at an early stage for a major investment deal. This, however, misunderstands the small business manager and the nature and ethos of the enterprise culture in the UK. There is much to be said for slower growth at the early stages, where outsiders' funds are not significantly involved, or, if they are, are not injected at a level to constitute a major risk. There is much to commend schemes and organizations which support entrepreneurs through these stages to develop products and acquire management skills. At that stage it becomes more feasible to assemble a strong, substantial business proposal requiring major investment.

An entrepreneurial venture capital organization (not necessarily a contradiction in terms!) can play a major role in supporting such small, technology firms whilst they develop their products, markets and track records. They will also have the benefit of getting into a potentially very high growth area at a very early stage; the risk will be that much higher, but the sums involved are relatively small.

The cost of assessment remains a problem, but only if the venture fund managers have to spend time looking at every project that comes their way. Competent Enterprise Agencies, Science Parks and Business Innovation Centres can play an important part here in acting as filters for appropriate venture funds, so that only the more promising projects are scrutinized by the fund managers themselves. This filtration would not, of course, abrogate the venture capital company's observance of due diligence on behalf of its shareholders, but it would reduce the number of total non-starters submitted to them in the first place.

Current progress

Some progress is being made. For instance, County NatWest Ventures (*see* p. 133) offers unsecured loans for technology-based firms in exchange for a minority equity stake.

The lack of investment support for technology-based firms has led some Science Parks to develop their own funds. The management company of Aston Science Park, Birmingham Technology (Venture Capital) Ltd., is the forerunner of these, the fund being aimed at companies within the City of Birmingham. It concentrates exclusively on technology companies, and offers investment between £10,000 to £250,000. It offers seedcorn capital, development and second-round financing. Other Science Parks have attracted seedcorn fund initiatives. A recent example is the new Oxford Science Park NatWest Seedcorn Fund which has been launched by the NatWest Technology Unit. This £3 million fund is designed to provide funds for use in either the generation of new products or product development in technology businesses.

The Kelvin Technology Department, involving the Scottish Development Agency, and the Universities of Glasgow and Strathclyde, also focuses on technology-based projects. Its average financing is quite small, at about £25,000.

Some universities are developing funds to support innovation emanating from their applied research work. University College

London has a fund involving the National Westminster Bank and UCL Ventures Ltd. The University of Birmingham Research Park Fund is a consortium involving SUMIT Equity Ventures and Barclays Bank, with Deloitte, Haskins and Sells offering management advice to the firms involved and Birmingham Research Park itself facilitating access to the technological expertise of a major university.

Other comparable schemes are in gestation. This kind of innovative thinking, born partly from frustration at the unwillingness of conventional sources of capital to assist in the development of technology-based businesses, is at last providing some limited access to risk capital. If your business is probably going to be located in or close to a Science Park, it would be worth finding out if the Park has such a scheme in operation.

The HATTSPI Scheme run by Hambros Advanced Technology Trust was an attempt on a national scale to bridge the gap. The scheme was jointly devised with the UK Science Park Association, as a source of seed funding principally for clients of Science Parks, with a maximum single investment of £100,000. The objective was to encourage the technology-based small businesses to work with science park managers to develop their business plans. The original fund has now closed, but a further fund is to be launched in 1992 (*see* p. 155).

■ APPROACHING THE VENTURE CAPITALIST

What they want from you

The problem for the technology-based entrepreneur is how to get a share of the enormous sum said to be available into his seemingly minuscule project. To see the best approach to take to achieve this, it is important for you to look at your project from *their* viewpoint. Essentially, their job is to *minimize risk* on behalf of their shareholders.

A first class project

The principal requirement of the venture capitalist is no different to that of the high street bank: a sound Business Plan is a prerequisite. They also expect a market-led product or process, with a clear 'edge' on existing ones, and a first-rate management team. This gives the potential for a high growth rate and a good market lead, thus justifying the risk and providing them with an 'exit route' after a relatively short time. In this way they can realize their investment and recycle the money into other projects.

A seat on the board

Venture capitalists see themselves as good at managing growth, and will normally wish to input their own expertise and connections to achieve this. Most will wish to nominate a non-executive to serve on the board of the companies in which they invest, and may even charge a fee for this. To some entrepreneurs, this is adding insult to injury: 'not only do they want 30% of the company, they also want to put their own man in to tell me how to run it and then charge me for the privilege!'

In practice, however, it can be very advantageous. For a start, investment organizations can take widely differing views of the role of their non-executive directors. Some adopt a 'hands-off' approach, and exert only a passive role. Others have a much more 'hands-on' attitude, and will involve themselves in many aspects of the company's direction, especially the financial ones.

This can in fact be a great help if you are developing an innovative company, because it is quite likely that the weakest area will be in financial management and expertise. Such a person, far from being a 'cuckoo in the nest', could well complement the management skills already there. As one entrepreneur in the control engineering area put it: 'When the presence of a non-executive director was first insisted upon, I felt resentful – but I needed the cash. I was then invited to discuss my Business Plan with one of the Managers [of the Venture Capital organization], and was introduced to a man who seemed just as entrepreneurial as me, but with a good deal more professional financial experience.

'It was he who joined the Board as a Non-executive Director, and his attitude has been generally positive, and invariably helpful. He understands the problems of the business, and I have frequently used him to bounce ideas off. So we not only got the cash, but we got the advice we needed to use it effectively.'

Of course, it is not always like this. There can be personality difficulties, and, if the business runs into trouble there could be an increased risk of friction. On balance, however, a business needing a substantial sum of money needs to have someone on the board who is accustomed to managing such sums. In such cases, a 'hands-on' investor is to be preferred.

Specific objectives

The needs of the various investment institutions vary substantially, although in general terms they will be looking for well-rounded management teams with market-led products and good growth potential. Whether or not they will be looking for a steady dividend flow depends on the nature of the fund.

The pension funds will look for a constant dividend flow from their investments, in order to support the regular payment of pensions. Venture capital organizations seek innovative, market-led, very-high-growth businesses, and look for a well defined exit route so that they can sell their investment and move on to others. The managers of BES funds manage the money of individual investors who are paying income tax at the higher rate, wish to take advantage of the income tax relief inherent in the BES and would prefer to see their investment grow over a period of several years, rather than receive a series of dividends.

Some of these objectives may match your own. For instance, if you do not expect to have a surfeit of cash in the early years, and you want what cash there is as working capital rather than being paid out as a dividend, then some of these schemes may be very attractive. Some, like the BES, usually deal in relatively small sums. Most, however, prefer investing substantial sums, since the costs of assessing a business by the fund management for a potential investment is roughly the same, whether an investment of £70,000 or £700,000 is contemplated.

So it is important to seek equity from the sort of fund whose objectives most nearly match your needs. It should be clear that, unless you have specific knowledge and expertise yourself, expert advice is necessary.

Flexibility

Fund managers prefer to structure the deal themselves, and are probably much more skilled at it than you. So do not be too rigid with your ideas, and be prepared to be flexible. Of course, if none of the offers appeals in the least, you have the opportunity to say no, and shop elsewhere.

Remember also that you and the investor *should* be on the same side!

Certain rights

In exchange for the investment, the investors will demand certain rights. One of these, the right to appoint a non-executive director, has already been discussed. Another may be the right to a dividend.

There are also rights to vote on specific issues which affect the corporate nature of the firm. These could include major policy changes, levels of borrowing, major acquisitions and disposals of assets, etc.

What you must watch

Check the details

The variety and complexity of equity deals means that you must be prepared to look carefully at the details, and negotiate with the investor. Equity deals are, in principle, flexible enough to be shaped to meet a variety of needs. Because of this, they can be difficult to follow. It could be a great advantage to seek advice from an experienced accountant or management consultant at this stage. They can cost a lot, but could save you much more in the long run.

Don't take the first offer

It is just possible that you may make your first contact with just the right investor, but it is unlikely. Different funds and their managers have widely varying preferences, and if your project does not appeal they are hardly likely to offer you an attractive deal.

So seek out those organizations which have interests parallel to yours, and shop around. If you find a good investor, you have someone who is prepared to share in your losses as well as your success. Such support is well worth cultivating.

Signpost the way out

The venture capitalist is looking for continued high growth from his investments, and does not want the fund to become locked up. Your Business Plan should thus offer an exit route.

■ SOURCES OF VENTURE CAPITAL

The British Venture Capital Association (BVCA) has over 120 full members. The problem is finding those which are appropriate for the entrepreneur starting or diversifying a technology-based business. The 1991 Directory of the BVCA includes a number of Members who might provide such a service, but there is no firm information on how well these criteria are observed in practice (frequently, as was noted earlier, they are not). In addition, quite a number of funds represent attempts by regional bodies to overcome the dearth of funds available outside the South East.

A complete list of BVCA Members can be obtained from the Association, and they have a database of venture capital organizations prepared by Price Waterhouse. For a fee (currently £23 including VAT) they will select a list of contacts matched to your business criteria. KPMG Peat Marwick McLintock also maintain a database of providers of venture capital.

Fund	Minimum investment	Location preference
Alan Patricof Associates	None	None
Alta Berkeley Associates	£100,000	UK, Europe
Barclays Venture Capital Unit	£100,000	UK
Barnes Thompson Management Services	£100,000	UK
Birmingham Technology Ltd[1]	£20,000	Birmingham
British Technology Group	£50,000	None
County NatWest Ventures, Ltd	£5,000 (seed)	None
Derbyshire Enterprise Board	£50,000	Derbyshire
Doncaster Enterprise Agency	£10,000	Yorks/Humber
Enterprise Equity (NI)	£20,000	N. Ireland
Future Start	up to £150,000	Deprived areas
Greater London Enterprise	£50,000	London area
Hambros Advanced Technology Trust plc	£50,000	UK
3i plc	None	UK
Industrial Development Board for Northern Ireland	None	N. Ireland
Korda and Company	None	UK, Europe
Lancashire Enterprises plc	£50,000	Lancashire
Lothian Enterprise Ltd	£10,000	Lothian area
Oxford Seedcorn Capital	£10,000	Oxford area
Prelude Technology Investments	£20,000	UK
Scottish Development Agency	£50,000	Scotland
Seed Capital Ltd	£5,000	Henley-on-Thames area
Sumit Equity Ventures Ltd[2]	£15,000	Birmingham
Tayside Enterprise Board	£25,000	Tayside
Thompson Clive & Partners	None	UK, US, France
UCL Ventures Ltd.	None	University Coll. London
Venture Founders Ltd.	£100,000	UK, US, Europe
Welsh Development Agency	None	Wales
Yorkshire Enterprise Ltd.	£50,000	Yorkshire, Humberside

Table 6.2 Venture capital for technology start-ups

Notes:
[1] Principally supports companies on or derived from the Aston Science Park.
[2] Sumit have a stated preference for mature companies, but the associated Birmingham Research Park Fund invests in new projects located on or close to the Birmingham Research Park.

Table 6.2 is a sub-set of the BVCA Membership selected from their information using the following criteria:

1. The Fund should be available to UK-based projects.
2. The Fund should either specifically state a technology preference, or there should be a number of investments in technology-based firms already in the portfolio.
3. There should be a stated minimum preferred investment of around £100,000 or less.
4. There should be either a stated preference for start-ups, or at least no stated aversion to start-ups.

The addresses of these funds are given in the Appendix, together with other relevant organizations.

Specific organizations dealing with technology

There are included in Table 6.1 a few organizations operating on a national basis which offer venture capital to technology-based firms at the start-up level. It is worth noting some of their characteristics and hearing what they have to say.

Investors in Industry (3i)
3i, as it is invariably known, is the UK's largest venture capital group, and was originally established by the Bank of England as long ago as 1945.

Unlike most venture capital companies, 3i sets no lower limit on the size of an investment, and in 1990-91 some 27 investments were made in units of £150,000 or less. Of course, all this was not in technology businesses. Another advantage is their nationwide network of 24 offices, which at least means they are prepared to come out of London: in 1990-91 just over half their investment finance went outside London and the South East.

They make it clear that for technology-based projects the emphasis is on management teams with general management skills rather than individual scientists. They pay particular attention to whether the project is an application of a proven technology or whether it has to be proved first; whether there is a market 'window' which makes the timing of the project critical; what the competition is likely to be; and how the product will be distributed and sold to the end user.

3i operate a Technology Programme, and have played an important part in the development of many technology-based businesses. They have invested in several areas of 'leading-edge' technology, such as diagnostic biosensors, pharmaceuticals from transgenic animals, software engineering tools and semiconductor test equipment. Another initiative worth noting is IMPEL, a joint venture between 3i, REL (a 3i subsidiary) and Imperial College, London, which is designed to exploit innovative ideas emanating from the College.

The British Technology Group (BTG)

The BTG is currently a self-financing public organization, but is being groomed, amidst some controversy, for privatization in the near future. It is the world leader in the management of Intellectual Property Rights, and concentrates its efforts towards licensing new scientific and engineering products to industry. It also provides finance for the development of new technology.

It has a rather different attitude from most other venture capital institutions, which reflects its history: it was formed in 1980 from the old National Enterprise Board and the National Research and Development Corporation. Perhaps its most well-known past successes have been the backing of the hovercraft concept and the antibiotic cephalosporin. Recent diversifications include support for a laser mastering system for monitoring compact disc manufacture, a 'dipstick' cholesterol assay and continuously variable transmissions.

For academic researchers with a potentially commercial idea, BTG's 'seedcorn' scheme, perhaps in conjunction with a SMART Award (*see* p. 120) could provide a useful source of money for a feasibility study. This allocates small amounts – up to £2,000 – to progress the work to the point where more substantial start-up finance can be contemplated.

The BTG has made the management of licensing and intellectual property rights its *forte*, and will take the responsibility of the effort and cost of patenting the inventions it selects. It can fund development and seek joint venture partners, both in the UK and worldwide, to exploit the invention. The licence income is shared with the inventor.

It spends heavily on the achieving of patents, and also on legally defending them, and has been accused of spending more of its resources on legal costs than it does in supporting other aspects of innovation.

The HATTSPI Scheme

The HATTSPI (Hambros Advanced Technology Trust Science Park Investments) scheme, managed by Hambros Advanced Technology Trust, was an attempt to bridge the 'venture gap'. It was developed in conjunction with the UK Science Park Association, as a source of seed funding for clients of Science Parks. A major benefit of the scheme was to bring the entrepreneur, for the first time, in contact with venture fund managers who were prepared to talk in their terms.

This fund is now closed, but a new fund, with comparable but not identical objectives, has been raised, and will be launched in 1992.

Future Start

This is a rather specialized fund, raised from British Telecom and managed by Hambros Advanced Technology Trust. It is directed solely at the deprived areas of the UK, such as the inner cities. It offers seed capital up to £150,000, and its investments include technology-based companies.

Corporate venturing

■ WHAT ARE THE BENEFITS?

Corporate venturing, which is still a relatively new phenomenon in the UK, is the investment of one company in another, usually by the taking of a minority equity holding. It is not dissimilar in principle to venture capital funding, but it does have some major distinctions in practice. The essential difference lies in the sharing of *business* risks and opportunities as well as the financial ones.

Typically, the normal objective is to marry the technological skills and flexibility of a small company with the financial and marketing muscle of a larger one. The most successful examples of corporate ventures have been where there is a complementarity of technology between the partners. More often than not the deal involves more than a straight cash investment. It frequently constitutes a manufacturing collaboration, with the larger firm making the product on behalf of the smaller one. Alternatively, it could be a marketing liaison, with the smaller firm exploiting the larger company's distribution network and sales outlets.

For the larger firm seeking to diversify into new markets or new technology, the process can be a low cost, low risk alternative to acquisition. If you are an entrepreneur, there are also some good reasons to at least consider this route.

Access to management expertise

A major company will have management experience which could well be of assistance to you, perhaps in financial control or personnel management. At a more basic level, you may be able to persuade the larger firm to give you some administrative support, which could help to reduce your overheads.

Access to technical expertise

Perhaps the reason for your approach to the larger firm was because they could exploit or market your technology-based product. Nevertheless, the larger company may well have technical expertise and equipment, access to which could be invaluable to your project.

Increased credibility

With the right partner, a strong, constructive relationship can ensue. A formal tie-up with a well-known firm can give you increased credibility with your market and suppliers, not to mention your bank manager.

Increased security

In times of trouble, banks and venture capitalists have a tendency to 'pull the plug'. This is because the link between you and them is purely a financial one. The corporate investor is less likely to do so because the connection between you involves not only money but people, markets and products.

Increased market or supply

You will not be picking a corporate partner with a pin, but will choose one who has perhaps supplied you with components or has shown an interest in buying your products. Quite possibly, the relationship has developed over a period of time. This adds to the increased security mentioned above by giving you security of a good supply of components or a market for your product.

■ WHAT ARE THE HAZARDS?

Loss of identity

This is the 'Jonah and the whale' phenomenon, implying that the larger company swallows up the smaller one, dominating its management and dictating its decisions. It can start by the larger firm persuading you to give up more of the equity than you really wanted, even to the extent of relinquishing voting control over what was your business. The next stage can be the gradual imposition of the larger firm's 'culture', including its management and administration procedures.

In reality, this is bad for the larger firm too, since a principal benefit of corporate venturing to them is the tie-in to the flexibility of a small, innovative company. For you, it could be the last straw, since you may well have set up on your own largely to escape the bureaucracy of big business.

Loss of further finance

There can be a clash between your aspirations and those of the corporate investor when you decide to expand further by seeking additional equity finance. On the one hand, they may not have further available funds to invest, but, on the other, they may be reluctant to see their stake in your company diminished by an additional investor.

Loss of an exit

Corporate venturing should not usually be a lifelong matter, nor should such agreements normally result in your takeover by the larger firm – 'Jonah swallowing the whale'. It is important that an exit route be available which is satisfactory to you, since you may wish to regain total autonomy or perhaps wish to diversify in other directions.

■ THE AGREEMENT

Corporate venture agreements are highly complex and legally technical documents, and you are well advised to seek a professional adviser to represent you: your putative partner certainly will.

The major matters to determine are finance and control. The size and stages of initial and any subsequent injections of capital, and what each stage will depend on must be clear. The question of control and autonomy must be settled and agreed.

As implied above, the termination clauses are also a vital matter. They are important to the larger firm, who does not want to get 'taken for a ride', but the manager of the smaller firm *must* be sure that his future options are not unduly restricted.

■ WHERE TO START

Contemplation of a corporate venture is a more complex matter than seeking a bank loan, and the 'corporate venturing scene' in the UK is still in a fairly embryonic stage. Nevertheless, many small firms have already found it an appropriate means of securing the additional finance they needed and the impression is that it is easier to achieve than obtaining conventional venture capital.

Many entrepreneurs will already know where to start. They have regularly bought components from a large supplier and have got to know the people at the top. Or they have had a positive reception from the manager of a large firm who are potential users or distributors of their new product. If you are in this position, then why not broach the subject – but make sure first of all that you

(*a*) trust the person and
(*b*) that he or his authorized superior signs a Secrecy Agreement on behalf of the larger firm (there is a typical specimen Agreement at the end of Chapter 3).

If you are not in the position of knowing a potential corporate investor, your local Training and Enterprise Council, Enterprise Agency, Innovation Centre or Science Park should be able to assist. So will offices of major accountancy practices and management consultants. Also on a national basis, the NEDO Corporate Venturing Centre, which was established in 1988 by the National Economic Development Office, may be able to help. This organization was privatized in 1989 and is at present unique in the UK.

The NEDO Corporate Venturing Centre publishes a register describing new and established ventures seeking partners. The entries appear in the form of anonymous profiles describing the product or technology, the history of the venture, future options and resource requirements.

Subscribers make enquiries to the Centre, which then informs the entrepreneur. Contact with the subscriber is at the entrepreneur's discretion.

Chapter 7
MANAGEMENT AND FINANCIAL CONTROL

Introduction

Once you have acquired the initial management team and the cash, you are ready to start. You will, presumably, have taken legal advice over incorporation and be one of the subscribers to and directors of a literally brand new company. At this stage, once any initial excitement has worn off, you will probably realize that your problems are just beginning. *Now* is the time that the strategy in the Business Plan must be put into practice, and a tactical plan developed.

It is therefore important to review some of the likely areas of concern, and make some general comments. Each of the major areas can be complex subjects in themselves, and are the subject of other textbooks. It is not necessary for you to become an expert in every area, but it is important to know

(*a*) the essentials and major dangers and
(*b*) where to get expert advice.

Premises

Should you start at home, under the proverbial railway arch or in a luxurious Technology Park? The answer depends on the type of business, but a general rule is don't pay out more money on rent unless you need to.

Working at home is not necessarily the best answer, but maybe you do not want a full unit of accommodation for a month or two, until you have assembled your staff and ordered necessary equipment. One answer may be the 'Rent-a-Desk' facility, offered by many managed workshops and Enterprise Centres on a temporary basis.

Most people like to start their new businesses close to where they live, and there are good reasons for this, not the least being the trouble and disruption entailed in moving house and family. However, remember that there are real financial advantages to be gained by locating in one of the Government's Assisted Areas. Chapter 5, p. 118 gives information on a range of grant assistance which is specific to these.

There are also clear advantages to the technology-based business to be in a location with other similar businesses, and with common services available. You may also need recourse to the R&D expertise available in the local university, college or centre of research. The growth and survival rates of small firms based on Science Parks and Innovation Centres are higher than those located elsewhere: the reasons may be complex – and include the proximity of a centre of technical excellence – but the facts seem clear enough.

It is important to take expert advice when entering into a leasing agreement. Your adviser should warn against committing yourself to a long-term lease which may be expensive to get out of. You should also think a year or two ahead, and acquire accommodation with the flexibility for expansion, if this is envisaged in your Business Plan.

I once again extol the virtues of Science Parks and Innovation Centres. Many of my current tenants started off in a single small office and have relocated within the Park several times as it became necessary. *They now occupy several thousands of sq.ft each and are still planning further expansion. Such flexibility seems eminently suitable to the technology-based firm with high growth potential.*

Personnel

Your staff are your greatest asset – and also the most expensive one. It is of cardinal importance that your human resources match the needs of the company. Staff need to be chosen carefully, employed under a proper legal basis with a clear job description and contract, and should be treated courteously and encouragingly. There are times when you will have to exert discipline. However, a carefully chosen staff *who are kept informed* will not only be an asset in the good times, but will show tolerance and support when things get difficult.

It is important to meet the legal requirements for employment. This is not difficult; there are excellent booklets available from the Department of Employment, and your adviser should also be able to assist. You will need to draw up Contracts of Employment and Job Descriptions. These may need special clauses in them about confidentiality, or restricting staff from working in a similar company for a given time after leaving. (Such clauses are not just for the R&D

people: to have a salesperson leave and take the client list to a competitor can be even more disastrous).

You may also need advice on pension and superannuation schemes, and other employee benefits such as sick leave, share options, profit-sharing and bonuses.

It is best to incorporate all this advice and the decisions made into a staffing and recruitment procedure. A coherent set of such procedures can greatly assist the entrepreneur 'free-up' more of his precious time for running the business, and can provide a platform of administration from which the company can grow into a coherent, supportive team.

Administration

There will inevitably be legal and administrative matters to deal with, ranging from PAYE and National Insurance for the staff, to COSHH Regulations, VAT, Customs and Excise and Social Security.

These are mandatory tasks, but ones which most managers find irksome. Perhaps the best advice is not to leave them on the desk – they won't go away – but again to define simple procedures. Decide what is to be done, who is going to do it and who will check to see that it has been done. Bring the whole of the management team into this, so they can share the tasks and help formulate the strategy.

The carrying out of the formal accounting procedures is also a necessary administrative task. It is separated here from the subject of financial *control*, since the exercise of the control is essentially a *management* responsibility and not necessarily that of the accountant or assistant who simply collects and collates the figures.

Nevertheless, the records do have to be kept and the books brought up to date, initially to provide management accounts and eventually for the statutory audit. The staff have to be paid, the statutory payroll matters have to be attended to. The debtors must be chased and the creditors paid. Regular management accounts must be prepared so that the management can do their job (*see* p. 166).

How this is achieved depends on the size of the company and the volume of transactions. For a small, start-up operation, a firm of accountants should be able to supply help on a contractual or part-time basis. As the company grows, it should employ an accountant or financial manager. The faster the growth, the more important it is to have regular management accounts.

Financial control

This book is aimed at serious innovators, that is, it assumes you want to be in business in order to make money for yourself and your investors. For this reason, the control and management of the finance is paramount.

Even if you have by chance 'struck gold', the establishment and growth of your business will be a critical time, and control will have to be exercised rigorously over all aspects of the company. Indeed, if you have 'struck gold', there is the threat that the demand for your product will outstrip the amount of cash available to meet that demand. This is the danger of 'overtrading', and there are plenty of firms that have gone out of business holding full order books. Thus, the control and management of the finance in your business is *the* most important aspect, and it is therefore given separate and fairly detailed discussion.

Management and control is, of course, important to all businesses. There are, however, important aspects for technology-based firms which need special care and emphasis.

Do not imagine that, in the eyes of your advisers and investors, well-managed businesses always run exactly according to the Business Plan. Perhaps it would be nice if they did, and it would take much of the risk out of starting a business (and also much of the challenge!). However, businesses are planned on the basis of the best available information, and even the most carefully considered can be blown off-course for a whole plethora of reasons, both positive and negative.

As examples, there are the more obvious internal things which can cause problems. The key manager can fall ill, the R&D phase can encounter totally unexpected difficulties or perhaps a competitor, having found out about your new product, has unexpectedly brought out a competing one or has immediately reduced the price of his product to undercut yours. Although some of these factors should have been considered in the Business Plan, the timing and extent of them cannot always be foreseen.

Then there are unexpected changes in the world or national economy which can quickly exert acute effects at the 'micro' level. How many businesses in 1989 (*and* the banks which encouraged them and lent them money) could have foreseen the downturn in the UK economy of 1990-91? What influence will the advent of the Single European Market have on the small technology-based firm? Will the countries of Eastern Europe now emerging from Communism constitute a threat as competitors or present an opportunity of a bigger market? There are many such uncertainties, and we live, as the Chinese curse says, 'in interesting times'.

It is unrealistic to expect the entrepreneur to consider all such factors in depth, unless they are of specific relevance to his business. The manager of a new business should be concerned primarily with the management of his finance and resources and with the market into which he is going to sell. These are exactly the aspects which the Business Plan addresses, which is why, in a changing world, the production of the Business Plan is still vital.

The business plan and reality

Why is so much stress laid on your Business Plan?

You drew it up to assess the potential viability of your business, and you have persuaded others to lend money on the strength of the Plan. The Plan's usefulness does not, or should not, end once the necessary investment has been acquired. The Business Plan is your blueprint for the development of the business, and you should measure the actual progress achieved against the Plan. It will then be possible to identify reasons for the deviations, and, if necessary, to take appropriate action.

The Business Plan will contain a budget and cash flow projection. The curve of monthly cash flow will typically be of the pattern shown in Fig. 7.1. Initial purchases, especially of capital items, plus the fact that sales will not begin until the product has been made and marketed, causes the first part of the curve to dip downwards to give a substantial outflow and cash deficit.

If things go to plan, sales begin to pick up and the negative gradient flattens out. The curve then turns upward again to eventually cross the axis; from then on the amount of cash received exceeds that

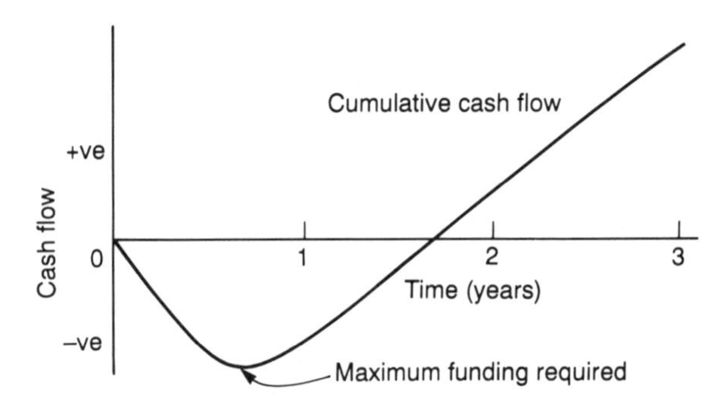

Fig. 7.1 Typical anticipated cash flow curve

spent, and the cash flow is positive. The first objective should be to minimize the length and depth of the negative part of the curve, or the 'valley of death', as it has been termed.

That at least is the theory, and many textbooks give the impression that this is just what should and will happen in any well-run business. In reality, in my experience, the picture is quite different. Deviations from the so-called 'ideal' are more common in technology-based businesses.

Frequently, the deviation from the Business Plan is negative. For instance, the research and development takes significantly longer than planned, the price of components increase, the market takes longer to penetrate or the economic climate becomes adverse.

Less frequently, however, the variance is positive. The market welcomes the new product or service, and rushes to buy it. One firm on my Science Park exceeded its sales forecast by a factor of 2.7 in its first year of trading. This may sound great news, and in one sense it is. However, the additional cash required to support this unexpected level of trading caused real problems: it is just as easy to develop cash-flow problems from overtrading as it is from insufficient sales.

Another common feature of technology-based firms stems from the need to capitalize on a newly-found market and develop further products. In Fig. 7.2, for instance, a company which has developed and marketed a new product finds that sales are going better or sooner than expected. At an early point it becomes clear that a further phase of development could produce a second product – or perhaps a range of products, for a much larger market. If they are to expand, and to be first into this large market, they will need to raise further loans or equity capital in order to pay for the development of the second product.

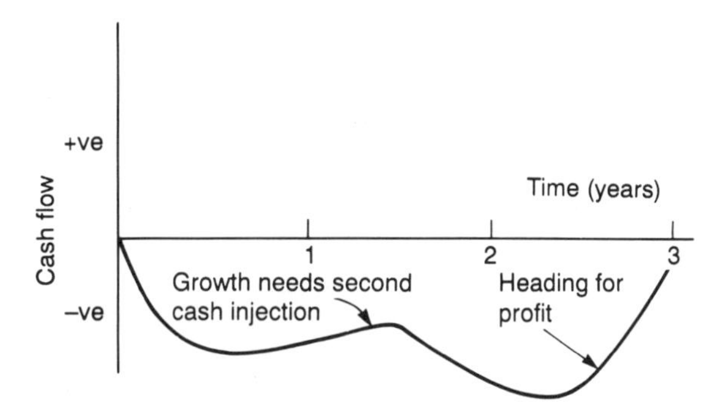

Fig. 7.2 Example of a cash flow curve for an expanding company

The cash curve thus take a nosedive again, perhaps even before it crosses the axis, and another, deeper and wider 'valley' has to be traversed before overall profitability results. Sometimes, two or three such phases of growth can take place, and the resulting company can be much bigger than the one originally envisaged.

From the viewpoint of those of us supporting business development, either by advice or investment, this is the best type of company to support. They have the potential for real growth, leading to significant benefits to the economy, employment and, ultimately, the investor. The new, small firm which stays small but profitable is entirely laudable, but it does not bring the added value of a firm with substantial growth potential.

Doesn't this kind of phenomenon therefore mean that the Business Plan, with all the hassle of constructing cash flows and so on, is really not worth the effort? Wouldn't it be better to test the market, and if people want to buy the product, go ahead and make it? If things are going really well, then all we need to do is to borrow more money in order to get more profit at the end.

Not a bit of it! In fact, the various scenarios described make the business planning exercise *more* important rather than less. It may seem obvious when put so baldly, but only if you know where you're going can you decide whether you are getting there or not.

Alice said 'Would you tell me please, which way I ought to go from here?' 'That depends a good deal on where you want to get to,' said the Cat. 'I don't much care where,' said Alice. 'Then it doesn't matter which way you go,' said the Cat. 'So long as I get *somewhere,*' Alice added as an explanation.

'Oh, you're sure to do that,' said the Cat, 'if you only walk long enough'.

Alice's Adventures in Wonderland, 1865

The way to avoid this kind of scenario is more planning. The Business Plan is still the 'touchstone' of the whole operation, but more detailed financial and project plans will need to be drawn up to fulfil it. These are not terribly difficult things to do, but they need the same mental disciplines and time that were required for the derivation of the Business Plan. The budgeting and financial plans may well, however, need the assistance of an accountant or financial adviser.

Financial information for management

Ignorance possibly may be bliss at times, but not if you're running a business. Information, and regularly updated information at that, is vital. Without it, you can't *know* what the true cost of a product is, or where you might best be able to cut down on expenditure. You, and, indeed, your whole management team need to have regular information on sales, the progress of work, the costs incurred and the state of your bank accounts.

■ ACCOUNTS SYSTEMS

This book will not concern itself with the details of setting up a suitable book-keeping and accounting system, since these details are not usually the prime concern of the technical innovator. There are also many books dealing with the subject, such as the companion volume *Book-keeping and Accounting* by Geoffrey Whitehead (Pitman Publishing, 1989). A suitable system will nevertheless have to be set up, and it must produce the records needed to comply with current legislation, particularly with regard to the presentation of year-end accounts. Suitable accounts or book-keeping staff (part-time or full-time) will have to be engaged.

From the viewpoint of management, it is important that this information in the various books and ledgers gets collected and collated into a regular series of financial 'snapshots' of the company's progress against budget.

The relatively low cost, ease of use and flexibility of software packages now available means that the accounts of all but the smallest and embryonic of firms are best established on a microcomputer. If you are just starting up, however, beware of going into the nearest supplier and buying the most expensive computer and sophisticated software. Take advice, and buy a system which meets your present needs and which can be upgraded a year or two hence, when the company's volume of transactions has increased.

Even the least expensive accounts software packages will have a facility to generate management reports with a long series of 'standard ratios' for accountants to pore over, but the real need is a package which will give *you* the figures needed to control *your* company. Although the automatic reporting facilities and many of the figures produced will undoubtedly prove of value, it is important to decide what else is needed to monitor the specific problems of your company. Some of the possibilities are dealt with below.

■ ACCOUNTS POLICY

From the outset, it is important to decide on a rational and consistent accounting *policy*. Costing and Pricing policy will be considered on

p. 173, but consistency and appropriateness of treatment of, for instance, R&D costs and assessment of work-in-progress are also apposite. The best approach for both is to be realistic and conservative. To be fair, it is sometimes difficult in a technology-based business to assess these items. Nevertheless, it is better for the management to have a prudent ongoing view of progress than to be told by the auditors at the end of the year that the estimated profits are far too high.

Work-in-progress should only be classified as such if there is a clear indication that the costs incurred can eventually be recovered. It is prudent to value it at the lower of the cost or net realizable value. No profit element should be included unless it is part of an ongoing, long-term contract, and only then if it is going well to plan.

It is better to write off expenditure on development as it occurs, rather than offset it against hoped-for orders. It is very easy to incautiously overinflate a company's profits in this way.

■ FINANCIAL STATEMENTS

The key performance indicators (*see* p. 180) are derived from the main statements of accounts, the Balance Sheet and the Profit and Loss Account. Their value is, however, largely historical, and the information must be backed up by the Cash Flow Forecast and other operational parameters such as the current Order Book and the Outstanding Debtors list. This Section reviews each of the more important statements from the viewpoint of providing information for 'hands-on' management.

Profit and Loss Account

The Profit and Loss Account is a picture of the company's overall trading and profitability over a given period; it is essentially an historical document. The information for it is derived from the general ledger: how much product was sold, for how much and what it cost to produce. It gives the *Gross Profit*, which is the sales income less the cost of the stock, and the *Net Profit*, which is the Gross Profit minus the overheads and other expenses.

This is useful information, but, as it normally appears in a set of annual accounts, it is pretty non-specific. It gives no details of trends, or of the profitability of particular products in a range. Neither does it tell you how well or how efficiently the company is performing.

Efficiency, so far as strategic issues are concerned, means the return on the capital invested in the company, measured as net profit as a percentage of the capital employed (*see* p. 180). It tells the investors how well you are performing with their money, which should always be significantly better than if it had been put into a bank deposit account. This ratio, the *Return on Capital Employed,* is the most important single indicator in the eyes of the institutional investors.

The Balance Sheet

The Balance Sheet is a 'snapshot' of the company's position at one precise time, a balance of the assets against the liabilities. Its format allows the reader to identify the sources of funds and their application, i.e., how they have been used. For instance, funds invested in a company could have come from share capital, retained profits, long-term or short-term loans, a bank overdraft and the creditors. (Examples are given in Table 4.4.)

Some of the profits retained may have been deposited in the bank, but the rest may have been spent on new capital equipment, a development programme or on components (stock) from which finished or partially finished product has been made.

The balance sheet provides some of the data which can be usefully employed in assessment of the efficiency of a business, and several examples of its usefulness are evident on pp. 180–182. However, the information can rapidly become outdated, unless the accounts software used has a program to generate a monthly balance sheet: many do have such a facility.

The Cash Flow Forecast

The Cash Flow Forecast is probably the most important strategic indicator for controlling the finances of a business. Do not confuse it with a monthly budget forecast of profit-and-loss. The Cash Flow Forecast takes into account the delays incurred in receiving cash due, so that the actual flow of cash in and out of the company is predicted. See Table 4.3 for a list of the principal factors which have to be considered.

Essentially, the Cash Flow Forecast measures the *liquidity* of a firm; the amount of cash actually available at any given time. It will be required in any Business Plan for investment (*see* Chapter 4 p. 101), but avoid the temptation to regard the matter as past and done with. Nothing could be further from the truth.

Recessions can bite within a month or two, and rapidly create a credit control problem which plays havoc with earlier forecasts. Or perhaps there is a two-month delay in getting the new product from the development phase and into production, so all your sales estimations and hence the eventual inflow of cash is delayed. Or there could be an unexpected demand for your product, which means the purchase of additional components and labour; the additional anticipated profit is excellent, but will the firm go broke before it ever sees any profit?

If you have not already done so, read and consider the examples given in Chapter 4 p. 104. This was devised to illustrate the importance of the Cash Flow Forecast in business planning, but it serves just as well in the context of day-to-day financial control.

Imagine *you* are the manager of the firm that produced the original forecast in Table 4.2 and Figure 4.1, and that things started slower than you had hoped. Because you are a good manager, you prepare a new forecast. This turns out to be nearer to that in Figure 4.2, where the sales income is deferred. It will be a shock to do the Cash Flow Forecast and realise that urgent action is essential, but it will be even worse *not* to do the forecast and find that the money has run out before you know it – literally!

The Cash Flow Forecast is therefore something which should be regularly updated. It is quite a task drawing it up first time, but the updating process should be somewhat easier. If your accountant or adviser helps, make sure the initial assumptions are sound and that you agree with them. It is too easy to superficially concur with a page of assumptions presented to you by an accountant and then let him put everything into a smart spreadsheet whose predictions everyone accepts without question.

There is no doubt that the microcomputer can help. There are some good programs which will convert a Trading Forecast into a Cash Flow Forecast by asking for additional factors such as the average time you take to pay your creditors and the time taken for your debtors to pay you. Such software will then compute the predicted inflow and outflow of cash, enabling you to decide whether and when further investment might be necessary. The other advantage of the computer is its ability to test any number of 'what-if?' options.

Be careful, however, since their ease of use can be deceptive. For instance, when forecasting a monthly income or expenditure level from an annual budget it is very easy to press the button to divide everything by 12, and therefore totally ignore the seasonal fluctuations which affect every business. If you are one of those people who rely on an accountant to do this sort of thing, be very careful to check not just the bottom line, but the top line and all the assumptions as well. If it is your business, it is you who must ensure 'due diligence' and take ultimate responsibility.

In smaller technology-based companies run by a technologist there is a natural drive to push ahead with the R&D, get the product out and wait for the sales income to arrive. It is important for someone, ideally within the company, to look at the Cash Flow Forecasts regularly to see if the company's position can be improved by judicious rescheduling of investment, repayments, etc. Considerable savings can result, and any additional cash in the company can always be put to work in an interest-earning account.

□ In 1991 *The Independent on Sunday* published a short article by Alan Toop, who, after working in marketing for many years for Unilever and J. Walter Thompson, set up The Sales Machine International, now one of Europe's top sales-promotion companies. The article on the early days of The Sales Machine, is one of a series entitled 'My Biggest Mistake', and is wonderfully germane to this context. Thanks are accorded to Alan Toop and *The Independent on Sunday* for permission to quote from it:

'My biggest mistake was to start my own company without having the slightest idea of the importance of cash flow...

'We got off to a brisk start... The budget for year one showed our target income, a detailed analysis of expenditure and predicted a small profit. I was good at this sort of budgeting exercise. In the past, I had been required to prepare them for big brands I was responsible for, and I was well-trained in controlling them tightly.

'Of course, I knew we would need some money to finance the company... and I had arranged all that with the bank. After four months, our income was on target and expenditure was as budgeted. Great.

'We did, however, seem to be using [the money] rather quickly. So, for the first time, I sat down quietly, wrote the next six months across the top of a sheet of paper, and under each month pencilled in how much money we were due to receive and how much we were due to pay out. I had in fact reinvented the cash flow forecast. Reinvented, because in the positions I had held in big corporations, cash was something you obtained by simply turning on a tap once your profit-and-loss budgets had been approved by the board.

'What this cash flow forecast demonstrated was that the cash was going to run out in three months... The company was going bust... For the first time, I realised the inadequacy of profit-and-loss budgets.

'In the event, we managed to salvage the company – just. All my time was concentrated on attempting to improve disastrous cash flow. Terms of business were rewritten, staged invoicing dates were built into all new contracts and hurriedly renegotiated on current projects...

'Overdue invoices were vigorously chased. Cash-flow forecasts became the key management tool of the business, and nothing was more important than updating them monthly.

'Have I learnt from this near-disastrous mistake? I hope so. We are planning to open our fourth overseas office in India at the end of this year... the key planning document we've still to finalize – and we won't set a day to open until we have – is the new company's cash-flow forecast.' □

Order book

The value of firm orders received will certainly be used to assess the immediate stock and production needs – and hence the need for cash. However, there needs also to be an assessment of likely orders together with an estimate of just how likely they are. This book has emphasized the desirability of conservative assessment of sales and work-in-progress, but it would be silly to ignore any planning needed to cope with a large potential order until after the order was formally signed.

The impact of securing a large order on the cash-flow projection can readily be assessed by using a microcomputer. As was indicated above (*see* p. 170), this is just the kind of occasion when evaluation of several possible scenarios becomes important.

Statement of debtors

A monthly statement of aged debts should be regularly prepared and used in conjunction with the system installed for chasing overdue payments. Make sure there *is* such a system: there is no reason why you should let other companies use your money for months on end.

Management should also use such a statement to check the efficiency of credit control, and to update the Cash Flow Forecast accordingly.

Key control areas

A common problem amongst all small businesses and a particular fault of technology-based ones, is simply not knowing what is going on, especially in terms of finance. Technology businesses are often managed by people who are highly competent, either technically or in production or marketing, but who lack the business experience to keep up with the finances.

If you have any suspicion that you might be in this category, then there are a number of things you need to get a grip of: costing, pricing and control of the working capital. Let us now look at these, and then see how a system to monitor and control finances on a day-by-day basis can be established.

■ COSTING

'How much does it cost to make and sell?' seems an easy question to answer; yet it can be surprisingly difficult. For technology-based projects, initial costings will have been done by 'back of envelope' calculations by technically competent and numerate entrepreneurs, who, having shown that the whole scheme is financially feasible, then move on to seemingly more important things.

For a rapidly expanding business, this is insufficient. Predicted costs need to be assessed and compared with actual values. The information will be important to pricing policy, especially in a competitive area where margins may be slender.

Costs can be subdivided into *direct* and *indirect*.

Direct costs are those costs incurred in actually making the product or providing the service. They include the costs of the raw materials and components, the production staff costs, the energy costs of production and the costs of packaging and distribution.

Indirect costs are those which are essentially invariant with the number of items produced or units of service provided. They embrace management staff costs, rent and rates, insurance premiums, interest charges and other fixed items of expenditure.

The total unit cost is the total of the direct and indirect costs divided by the number of units sold. In accountant's jargon, assessing costs on this basis is known as *absorption costing*, because it absorbs the total costs into each unit sold.

There is, however, another useful way, called *marginal costing*, which only includes the direct costs in the computation. Although this method dodges the allocation of indirect costs, it can prove a useful guideline on many occasions.

For instance, suppose you produce a product whose total costs (absorption costing) amount to £100, but whose marginal costs are £80. Suppose then that you are negotiating with a customer who wants you to supply the product for £90. Is it worth it?

The obvious answer is 'no', since you would be selling the item for less than the total costs involved. However, if your company is operating at less than full capacity, it *would* be worth it, since every extra items sold at £90 would not only cover the direct costs but would contribute an additional £10 (£90 – £80) to the indirect costs. The detailed knowledge of the costs involved can thus be of substantial importance.

The details of systems for collecting data to enable costs to be assessed and of the complexity of the assessment is more a matter for a cost accountant than the entrepreneur. If you have accountancy skills, then fine; if not, it is better to employ someone to set up the system and do the calculations for you. Nevertheless, it is important to understand the principles of costing and to know which questions to ask. If you are the manager, it is your decision.

Questions about costs are also pertinent to the consideration of the 'right' price for the product, which is discussed next.

■ PRICING

The 'right price'

There is no infallible guide to the 'right price' for a product. It will depend on the fluctuating price of components, the state of the economy, the competition and the market's perception of its value.

The simplest way is to add up the costs of the components and labour, stick on a certain amount for overheads and then add a certain profit element. This is known as 'cost-plus pricing' and its limitations soon show up when it is applied to a range of products, where the cost per product depends in a non-linear fashion on the total number of units being produced. The overhead element is vital, especially in technological products, where the costs of research and development must be recouped and enough profit acquired to support the next phase of R&D. The main problem with this method, however, is that it ignores the market place.

An alternative is to take cognizance of the market, and base your prices on what competitors charge. Indeed, one should always begin and end any discussion on prices by referring to the market place. This is more difficult with an innovative product, and it may be necessary to consider what the customers might be prepared to pay for something they would probably like but do not have already. If your product is unique, there is a possibility that they would be ready to pay a high price. So, charge them a high price, but be prepared to lower it once the competition emerges.

There is an inherent tendency for new businesses to underprice their products or services, in order to generate sales and income rapidly. This is understandable, and in many instances will be necessary to woo customers away from the competition. However, if you have something which is unique and attractive to customers, be prepared to charge a premium. If the product develops a good 'image' and attracts a good share of the market, the price should be kept high for as along as possible; but don't forget that a high price will also hasten the launch of competing products at lower prices.

On this matter of 'image' and pricing, it always amazes me how one can put fizzy water in a moderately shapely pale green bottle and sell it at what must be an enormous mark-up. I can only admire the success of the marketing campaign which has given this product such an attractive image.

Break-even analysis

Your Business Plan will have assessed the actual cost of production of your product or the cost of delivery of your service. Such assessment will at first be done on the basis of the *variable costs* of the business. The variable costs are those costs which directly fluctuate with the amount sold. Typically, they include costs of raw materials, direct labour, energy used for production and delivery costs.

The business also incurs substantial *fixed costs*, which are those which are incurred independent of the level of production or sales. These include rent, rates, administration costs, depreciation, and electricity which is not used in the actual production process.

Never mind here about allocating them. The point is that your sales have to cover both the variable and fixed costs, otherwise you will never make a profit. The price on every article or service provided has to cover:

(1) the variable costs involved in its production and sale
(2) a proportion of the total fixed costs.

If these elements are included in the price, the effect of trading is to offset all the variable costs inherent in each article plus a proportion of the fixed costs. At some particular level of sales, *all* the fixed costs will have been recovered. This is called the *break-even* point (Fig. 7.3).

Above this sales level, the costs incurred are only the variable ones, so what was the fixed cost element simply becomes profit. It is normal, of course, to aim for sales well above the break-even point, so that a substantial profit is realized.

The calculation of the break-even point is conceptually very easy. The selling price of each unit minus the fixed cost per unit gives the value of the contribution to profit of each unit. The number of units needed to break even is then the total fixed costs divided by the contribution per unit.

Once the contribution to fixed costs from the sale of each unit is known, it is easy to calculate the profit you can anticipate from any

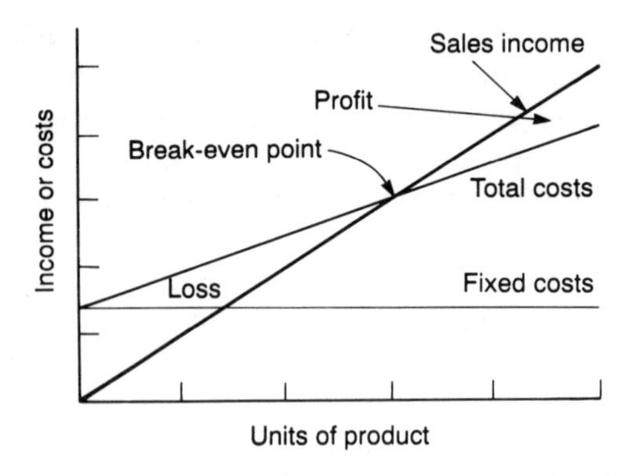

Fig. 7.3 Break-even point

level of sales. A typical computer spreadsheet application can be very useful to evaluate various 'what-if?' scenarios, employing different pricing and levels of sales, perhaps for different mixes of a number of products.

Although the calculation itself is easy, a problem is that it is just as easy to underestimate the true total cost of production. Usually, a number of fixed costs are either omitted or forgotten. Remember, then, to include things like depreciation, insurance premiums and accountant's and audit fees. They are all part of the cost the company has to cover by selling its products or services.

This type of analysis will soon show up which products and services contribute most to overall profitability, and can guide management decisions on prices, market segment and which products to promote. Any sudden change in variable costs can be quickly translated into a new break-even point and hence a new sales target.

■ WORKING CAPITAL

The working capital is the money in the business which is actually put to use in making the business work. Table 4.4 represents a simple Balance Sheet for a small firm, showing the assets of the business and how they are funded, and the working capital is the difference between the current assets (stock plus debtors plus cash) and current liabilities (creditors).

The Balance Sheet, however, is only a snapshot in time (*see* p. 169), and time is an important factor in the efficiency of use of working capital. Consider the sequence of events.

Components and raw materials have to be purchased, and the fabrication of the product means payment of salaries and overheads. The product has to be marketed and sold, and, while the selling task is in progress, it will be necessary to purchase more components for further production to begin. The suppliers of these will want payment for the first consignment, and so there will be a substantial outflow of cash before the inflow begins from the customers of the product. No doubt they will take advantage of the maximum credit period which is allowed.

Once this cycle is established, the inflow should be greater than the outflow because of the profit element, but the amount of working capital involved should be essentially the same, for the same amount of production. Any increase in production (or, for a service company, the turnover per unit time) will require further working capital, and part of the profit element is usually a significant source of this. It is estimated that, even for a well managed production company, an increase in turnover of £100,000 requires some £27,000 working capital (*The Genghis Khan Guide to Business*: Warnes, B. Osmosis Publications, 1984).

This means that a substantial amount of money is tied up in working capital in a typical business, and the harder it can be made to work the better. Time comes in here: the shorter the cycle time the better, since the ratio of working capital to profits is increased. There is therefore a natural desire to squeeze as much profit from as little working capital in as short a time as possible.

However, there is a danger, and one which frequently applies to growing technology-based companies. The attempt to grow quickly by using as little additional capital as possible can easily lead to running out of funds. The business can be selling well, and its profit-and-loss accounts show a reasonable profit. But profit-and-loss accounts are based on invoiced sales and work-in-progress, and not on actual cash received. If there are such cash-flow problems, the profits will be sucked into the working capital and, when these are exhausted, the business fails – despite being profitable on paper.

The incorporation of working capital needs into the Business Plan is thus vital. Much of it will be borrowed, especially in the early stages. This should cause no problem if the level, timescale and interest rates are known. The key to it all is the Cash Flow Forecast (*see* p. 169). Not *a* Cash Flow Forecast, but regular, ongoing, updated forecasts, comparing prediction with actual results, and based on a credible set of initial assumptions.

Controlling working capital

What means are there for controlling and optimizing the efficiency of use of the capital employed in your business? Most of them are fairly obvious, but should still be stated and considered. It is surprising how often the simple ways to improve matters are ignored.

Control of supplies

Why buy six months' supply of components or raw materials when two months would do? Careful purchase control with guaranteed delivery dates will help to keep the stock levels at an acceptable level. (There may be other factors here, such as big discounts offered for large orders – but you might still bargain to arrange scheduled payments over several months.) It can also be worthwhile for you or whoever does the purchasing to get to know regular suppliers, and place regular advance orders; this can be mutually beneficial, because your supplier can then plan his production or delivery schedules better.

Staffing levels

Watch the staffing level. Staff costs are one of the largest items of expenditure for most companies. Consider employing people part-time where appropriate: if the business expands, they can then be taken on full-time.

Personal and company expenses

Look too at your own expenditure. If you don't need a costly personal assistant, but can make do by sharing a competent secretary with the marketing manager, do you also really need such an expensive car? Of course, you must give the right 'image' to your clients and they deserve to be treated courteously, but are you going 'over the top' with expenses?

Remember that you are paying a high rate of interest on the money borrowed, and that that same money could be used directly in helping the firm to grow. Remember too that any large company that is a potential client will not be impressed by such lavishness, although no doubt they'll enjoy it at the time. They know that, if they order from you, the cost of it will be put on to their bill eventually.

Credit control

It is very important to set up and exercise a system of credit control. This area is often overlooked by the entrepreneur with a drive to go out and sell, but it is *essential*. It will cost money, but it can save more in the medium and long term.

Decide what your terms of credit are, and stick to them. By all means offer incentives for prompt payment if you think it will help in your line of business, but also have a system of sending out

Statements to debtors when they are 10 and 20 days overdue, and then put the matter in the hands of a solicitor. And be prepared to take defaulters to court if necessary. After one or two such incidents, word will get around that you are prepared to carry out your threats and there will be an increased tendency for bills to be settled on time.

Also be prepared to check on the creditworthiness of a company. You will find that others will check on you if you are a new, small firm – so you will be wise to ask for references to accompany a large order from a firm you don't know well. You are perfectly entitled to ask, in such circumstances, for some money up-front or for staged payments to be made during the course of the work.

There has been much concern about the time taken by firms to settle their debts, and unfortunately some of the very largest companies are signally failing to set an example. What may be an insignificant improvement in a large company's cash balance may prove the final straw for a small firm, struggling to maintain their viability with high interest rates. It has been good recently to see the CBI encouraging the larger firms to set an example by paying on time.

It is difficult for a small firm in this situation, and the hope often is that if further regular work can be generated, the cash will eventually come in regularly. My own experience suggests that it is important to make a nuisance of yourself to the large firm's accounts department. Find out who authorizes payment and signs the cheques, and then keep at them with letters and phone calls.

Nevertheless, don't be in the position of always blaming the other firm for not paying. Make sure you have a system which promptly informs your accounts when an order is completed and ready for invoicing, and that the invoices are dispatched at regular and frequent intervals. Ensure that there is an aged debtor list which is kept up-to-date. Who controls the list? Do you ever look at it? Do you ever take the trouble to phone one or two of the miscreants yourself, especially the ones you know personally?

Payment of creditors

Finally, it would be wrong if you expect your debtors to pay promptly if you do not yourself pay your creditors at the proper time. By all means reserve payment until the legally agreed maximum, but taking unfair advantage of stretching periods of credit only exacerbates someone else's problems and they too will be tempted to follow the same line. In the long run this benefits nobody. It is far better to pay on time, and develop a good reputation with your suppliers. This will be of help when you may wish to negotiate a discount or if, for once, you really do need to defer a payment for a week or two.

Performance indicators

Whilst it is natural for the entrepreneur to be concerned with the sales figures and ultimately 'the bottom line', the net profit generated, it is just as important to use financial and other data to assess and optimize the *efficiency* of this generation of profit. Such ratios or indicators of performance are much more valuable in assessing the use to which the management is putting their investors' money and their staff time. Some of the more important ones are briefly described below: for more detailed consideration, see any textbook on accounting.

■ THE CURRENT RATIO

This is the ratio of the company's current assets to current liabilities, and is obtained from the balance sheet. The current assets are the stock, prepayments, the debtors and the cash in the bank or in hand. Current liabilities are the creditors, accrued expenses (like the phone bill which is still to be received, but you know it's due soon), overdraft or short-term loans, and tax liabilities.

What the current ratio measures is the *liquidity* of a business, how readily it could theoretically pay all its debts in one go if it had to. From the creditors' viewpoint, it indicates how well they are protected by the available short-term assets.

A current ratio of more than 1.0 thus means that a firm is fully solvent. A ratio of much more than 1.0 means that there is spare cash around which is not earning, which is therefore inefficient. It is better to reinvest this to generate additional profits.

There is also a ratio called the *liquidity ratio* or 'acid test', which recognizes the fact that some assets take much longer to realize than others, and excludes stocks of raw materials and work-in-progress.

■ RETURN ON CAPITAL EMPLOYED (ROCE)

This is the ratio of net profit before tax (from the profit-and-loss account) to the capital employed to generate it (from the balance sheet), usually expressed as a percentage. It effectively measures the rate of interest from investing in the company. Your investors (including you, since you almost certainly have put your own money into your company) will all regard this as a vital statistic. They want to be sure that their investment in your firm gives a better rate of interest than if it had been put in a building society – or in someone else's business.

If the ROCE begins to go down, the cause may often be revealed by looking at other ratios.

For instance, examination of the profit to sales ratio can show whether those additional sales you were so pleased about *really* contributed to unit profitability. The ratios of profit to working capital, sales to capital employed, sales to overheads and so on, can all be employed as diagnostic indicators.

■ GEARING RATIO

The gearing is the ratio of the borrowed funds to the shareholders' funds (*see* Chapter 6 p. 135). It is important to secure the right mix of funds for a company, matching the type of finance to its purpose. If there is too high a level of external borrowing, there will be an increased burden on the cash flow because of the higher repayments plus interest.

On the other hand, many technology-based firms are highly geared of necessity, since the costs of initial R&D or of the purchase of expensive capital equipment may be well beyond the capability of the entrepreneur. In such cases, it may be preferable to seek venture finance with the consequential loss of some of the equity (*see* Chapter 6, p. 138).

■ DEBTORS' TURNOVER

How long on average do your debtors take to pay their bills? Are they using *your* money to help keep *their* business going?

Your accounts should be able to assess this, since it is easy to take the value of the debtors for the year to date, and divide it by the sales for the same period. This yields *the average time it takes for a debtor to pay up.*

$$\frac{\text{Debtors}}{\text{Sales}} \times 365 = \text{Average debtor days for year}$$

This ratio, if produced monthly, very quickly gives an indication of any slippage of the efficiency of credit control, and will enable targets to be set for the staff concerned.

■ OTHER RATIOS

There are numerous other performance figures which could be

listed. However, the best guide is to think out for yourself (or with your accountant) the critical factors specific for *your* business.

For instance, a consultancy firm will want to assess the cost-effectiveness of its staff by comparing the time they spend on chargeable work with that spent on administration, marketing, training, etc. It could also be important to know the backlog of work they have in hand, i.e., how long they can keep going on fee-earning work if no new assignments are secured.

A small manufacturing firm will wish to know whether components are sitting on shelves for too long before they are incorporated into equipment which is sold. By dividing the cost of sales by the annual average stock figure, the number of times the stock turns over in a year can be calculated. The more frequently it does, the more profit which is generated.

The company will also need to collect and collate information on failed components or failures of its product in use. Were most of the failures due to one particular component or to faulty production technique? Are the failures decreasing with time? It is easy to construct ratios which will indicate the changes with time in quality control rejections and customer returns.

Specific control points

The particularly critical areas will vary from company to company. For instance, there will not be the same need to exercise stock control for a consultancy firm as in a manufacturing company (although it may still be necessary to keep a watch on the stationery cupboard!). On the other hand, consultants will wish to monitor the proportions of chargeable and non-chargeable time.

Some common factors and how they are assessed are considered below. In every case, it is not enough to simply take one value of a parameter as a 'snapshot'. The kinetics, viz, the *change* in these parameters with time, is the important thing.

■ STAFF COSTS

These are usually the biggest regular outlay for a company, and it is important to ensure maximum efficiency. The number of staff productive hours, whether they are producing a product or a service, can be assessed from salary records and time sheets. The staff cost per item can also be computed from these data plus production records. Remember, however, that unfinished, rejected or returned items can quickly add to the true cost of production. There needs to be control over *quality* as well as finance. Any work inefficiently or incorrectly carried out imposes an additional cost on the company.

One simple ratio which can act as a rough guide is the sales per employee. Is this ratio, taken over the last few months, quarters or years, changing? Why?

■ RESEARCH AND DEVELOPMENT

Assessing staff efficiency in R&D is more difficult because of the frequently 'open ended' nature of the work. The actual costs however, can be computed and compared against budget; it is important to do this, since they can quickly escalate.

Insist on regular progress reports from project leaders. Try to set realistic quantitative targets with the staff concerned, so that they themselves have an achievable set of parameters against which their progress can be judged. Chapter 8 deals with project management in more detail.

■ PRICES

Pricing has been considered on p. 174. Develop a policy, and keep prices under regular review.

Internal factors such as production costs, and external factors, such as price-cutting by the competition, can rapidly affect sales and profitability. Prices may well have to be modified to meet these challenges, but don't keep on cutting them until the whole business becomes unprofitable. Use your original Business Plan as the guideline.

To meet special circumstances, it may be necessary to impose marginal costing (*see* p. 173) on some sales. If this is so, control it by comparing the actual sales or charges with those in the Business Plan. In this way, it is possible to identify the income lost by your decision.

■ SALES

Are the sales really coming in as fast as you would like to think? A system of formal Sales Orders and Sales Acknowledgements is the only way to properly estimate sales. A true sale implies a formal contract between the buyer and the vendor. Counting verbal promises or generally encouraging noises from a client as a sale are not acceptable.

Naturally, such encouragement should be treated enthusiastically, but in estimating the true progress of your business, take a prudent view.

■ PURCHASES

Once again, the budget must be the guide. A manufacturing company can, by arranging bulk purchases, reduce costs considerably. However, do remember that the price of some items can *decrease* as well as increase; buying a large quantity at a small discount may not be worthwhile in the longer term. Too much stock sitting on shelves means working capital tied up for long periods.

■ WORK-IN-PROGRESS

It is easy to be lulled into a false sense of security by securing a full order book, and then assume that the work will be completed on time. Regular monthly information on work-in-progress is necessary to monitor this, the *turnover* of work-in-progress.

The desirability of a conservative policy on work-in-progress has been discussed earlier (*see* p. 168).

■ DEBTORS

Credit control is one of the most important areas of financial control for any company. *Remember that a sale is not complete until the purchaser's money is in your company's bank account.*

The more adverse the economic climate, the more acute this problem becomes. Poorly managed firms can find themselves faced with a cash flow crisis, and the easiest way out of it is simply to stop paying the debts. Better companies can still run into cash flow problems in times of recession, but good financial management information can give early warning of impending difficulties.

■ OVERHEADS

Overheads have a habit of increasing by stealth. So keep a check against budget. Look at the changes in each category of overhead and for each person responsible: it may be you!

Controlling growth

The management of corporate growth, particularly of rapid growth, is a separate topic. Nevertheless, it is the customary objective of most technology-based firms, and, with good management (and just a little good fortune), you will find yourself facing the problems of expansion in the near future. There are some comments on the entrepreneur and the growing company in Chapter 1 p. 9.

Another volume in this series, 'Managing Growth' by Maureen Bennett (Pitman Publishing, London, 1989), deals in detail and clarity with the challenge of a growing business. From the viewpoint of financial control, however, the earliest need is to match the system to the growth. This problem will appear well before consideration of increasing equity and capitalization.

The increase in the product range, the number of sub-departments or cost centres and simply the increase in the number of transactions and the associated paperwork will require a more sophisticated and flexible accounting system. No longer will you be able to call an external accountant in once or twice a month to 'do the books': you will need to take on specialist financial and other staff.

Nevertheless, despite the increase in complexity, the same guidelines and precepts described in this Chapter hold true. It will be ever more necessary to ensure regular management accounts, and the need for positive financial control will be even greater.

Chapter 8

PROJECT PLANNING

Introduction

The inherent challenges of technology-based businesses means that special care and thought should be given to planning and control of projects. The need for financial control has been dealt with in Chapter 7, and this is true for any business, but there is more to developing a technologically complex product or service than managing the finances.

It is not just the difficulties in researching and developing a sophisticated product. There is a need to integrate the product development with the marketing and sales plan. It would be highly undesirable to plan a market launch to find that the product was not fully ready: the danger then is that the product will be sold without all the 'bugs ironed out' and your first and 'trailblazing' customers (*see* Chapter 9 p. 204) are sold a new product with a high failure rate.

At the other extreme, it would be foolish to spend a large sum on developing and producing substantial amounts of a new product for it to sit on shelves for months until the marketing and sales campaign gets under way.

Then there is the need to integrate the efforts of several people or groups, each with their own special skills and demands. Although most of this Chapter is concerned with planning of time and resources in a project, the human side of project management must not be neglected. Highly skilled staff need to be told why their expertise is needed, what the overall objective is and how long the work is expected to take. In other words, they need to be *motivated*.

They will work much better if they understand what the whole project is about, and the reason for tight timescales. People who are kept in the dark, given inexplicable tasks and impossible deadlines rapidly lose enthusiasm.

The ultimate objective with any business project is to satisfy the customers and the investors. Projects should therefore be planned carefully to be *within time and within budget*.

The elements of project planning

Project planning requires the management and integration of:

- **Time**. Any time constraints must be identified.
- **Activities**. The project should be broken down into a number of discrete and achievable events.
- **Resources**. This element includes people, consumables and overheads. Each Activity will have a resource implication.
- **Costs**. Similarly, the Resources involved will each have a cost implication linked to them. These must all be known.

The *objectives* of a project plan are:

- to assess the feasibility of the programme in terms of credibility and cost.
- to provide a means of ongoing assessment and control of progress.

As with the preparation of a Business Plan (Chapter 4), project planning is an iterative process. A first guess will almost certainly result in the project costing too much, the resources not being available or being duplicated, or the timescale not being appropriate. Several iterations may be needed before an acceptable compromise between time and cost is reached – and a compromise is usually the result.

Again, like overall business planning and budgeting, the control is exercised by comparison of actual with predicted position on given dates. Regular check-points should be built into the programme, when project reviews can take place.

■ DEFINE THE OBJECTIVES

This is the first step. Make sure that you *and your whole project team* know what the objectives are. The overall aim of the exercise and the time constraints should be clear.

Also be as quantitative as possible. For instance, if your objective is to develop a prototype immunoassay into a commercial diagnostic ELISA kit for use in the pathology laboratory, there are a number of essential numerical parameters which should be specified from the outset. These might include the sensitivity and reproducibility of the test, the maximum level of cross-reactivity, the allowable coefficient of variation within and between tests (microtitre plates), and the shelf life of the kit under normal storage and transportation.

■ LIST THE ACTIVITIES

Once the overall project has been defined, it should be possible to break it down into a sequence of sub-tasks or activities. Each of these may constitute a 'mini-project' with its own set of objectives. Indeed, a very large project is best considered as a set of interlinked projects. The linkage is essential, since

(*a*) the commencement of one project will depend on the completion of another and
(*b*) there may be common resources involved.

As an example, consider one of the largest engineering projects of modern times, the Channel Tunnel – a formidable planning task if ever there was one. The boring of the tunnels and the laying of the track are each immensely complex planning tasks in themselves. But one thing is for sure: the track-laying in the tunnels could not begin until the tunnelling was finished. On the other hand, the specialized rolling stock could be designed and built without the other two tasks being completed.

Each activity will have its own discrete start and end-point, but many will depend on each other, as the example in the previous paragraph illustrates. The sequential listing of activities will begin to give the programme a logical shape, and may indicate the specially critical nature of some activities. Note these early on, since their commencement or completion may well be the best time to programme-in project reviews and assessments of progress.

The rough initial sequence should now be considered in more detail to produce a Project Schedule. The major activities should each be considered in turn, relating them to their precursor and successor activities. Some might be fitted in at any time or within a given part of the Schedule. Others *must* be done in a particular order, and the sequence of these critical activities forms the *critical path* of the project. Any extension of the duration of an activity in the critical path will lengthen the time of the project overall.

It is prudent therefore to be realistic rather than optimistic about the duration of critical activities. Allow for the slight hold-ups which frequently occur. It is better to do this from the outset than be faced with possible claims for breach of contract by not, for instance, providing the completed prototype by a specified date.

■ ASSIGN THE RESOURCES

Each of the activities should then have appropriate resources assigned to it, including a manager who is responsible for keeping it to schedule and to budget.

Divide the resource element into people and 'things' (purchases, equipment and consumables). Each of the resources will have a price-tag associated with it. The cost per day or week of each staff member will be known, and the cost of consumables and supplies can also be estimated. Remember too that the use of perhaps expensive capital equipment or laboratory facilities should also be costed.

Working out the price per unit of use for each resource is sometimes tedious, but it is not difficult. It becomes more difficult to estimate the total use of each element; the tendency is to be over optimistic.

People

The staff element is likely to be the most expensive one, so be especially careful over the cost implication. Up to a limit, it is possible to shorten a task by increasing the staffing allocated to it, so, if time rather than cost is of the essence, it may be appropriate to increase the staff involved. However, there comes a point where increasing staffing any further yields rapidly diminishing returns: each individual's sub-task becomes trivial, there are too many people to manage at one time and they get in each other's way.

There is a temptation when considering project plans to treat people like objects, and just move them around in the plan without thought. Beware of setting up a tightly constructed plan and then finding that half the staff had previously booked holidays, or that you had agreed to send the relevant manager on a training course at the time he is needed.

Equipment, consumables and purchases

The project will need equipment and supplies, and it can be more difficult to accurately predict the amount rather than the nature of the resources needed, especially in an R&D programme.

It is almost commonplace to read in the newspapers of major engineering or defence projects which have grossly overrun their budget estimates, their project time, or both. Maybe there are good reasons for this which have little to do with good business practice and more to do with other factors. It is important, however, that the owner of a small business does not fool himself into bidding for a contract at a price and within a timescale which he has no earthly chance of achieving.

There is always uncertainty and risk associated with any project, and it makes sense to work out as realistically as possible what the best cost estimate is, and then add on a contingency cost on top.

It makes life somewhat easier if you have had previous experience of a similar project, and, even if you haven't, some of your staff may have undertaken *parts* of comparable projects before: their advice will be useful. It is advisable to discuss each task with the relevant staff at the planning stage.

■ SCHEDULE THE PROJECT

During the previous stages you will gain an increasingly firm idea of the *sequence* of activities: their order, and what the principal dependencies are. Now is the time to formalize this into a sequential and timed plan.

The optimum sequence

In some instances the timing will be unclear and perhaps not of crucial importance to the development of the project. This is so for some longer-term R&D work, which may contain certain speculative aspects of uncertain duration. However, to avoid cost overrun, it is important to give maximum times and costs to each activity, even for such programmes.

All sorts of excuses about being 'totally unable to predict progress' will frequently be aired by R&D groups. At project reviews, lack of progress can be covered by 'totally unforeseen difficulties' and 'another few weeks should give us the breakthrough we need'. Another favourite is 'we haven't spent the time following up the original aim. We have been following a totally new line of work, because an early experiment revealed a fascinating new angle which has got all kinds of new possibilities.'

So it might, but the organization is spending money on a totally speculative new area unknown to the management and, at the same time, the development is diverted away from the original goal, the new product.

Since I have both participated in R&D and managed quite a few such programmes as well, I have both sympathy with the impossible demands sometimes made of R&D groups and knowledge of the inventions employed by them to squeeze a few more months' cash out of the management. This book is written for managers, and if you have the task of managing or being responsible for an R&D group, do keep them closely to agreed objectives and try to hold them to some sort of time scale.

The sequence of events, without timescale, can be depicted graphically in a PERT chart (Project Evaluation Review Technique). An example is given in Fig. 8.1. This type of formalization is extremely useful, and it is well to begin doing this sort of thing early on with

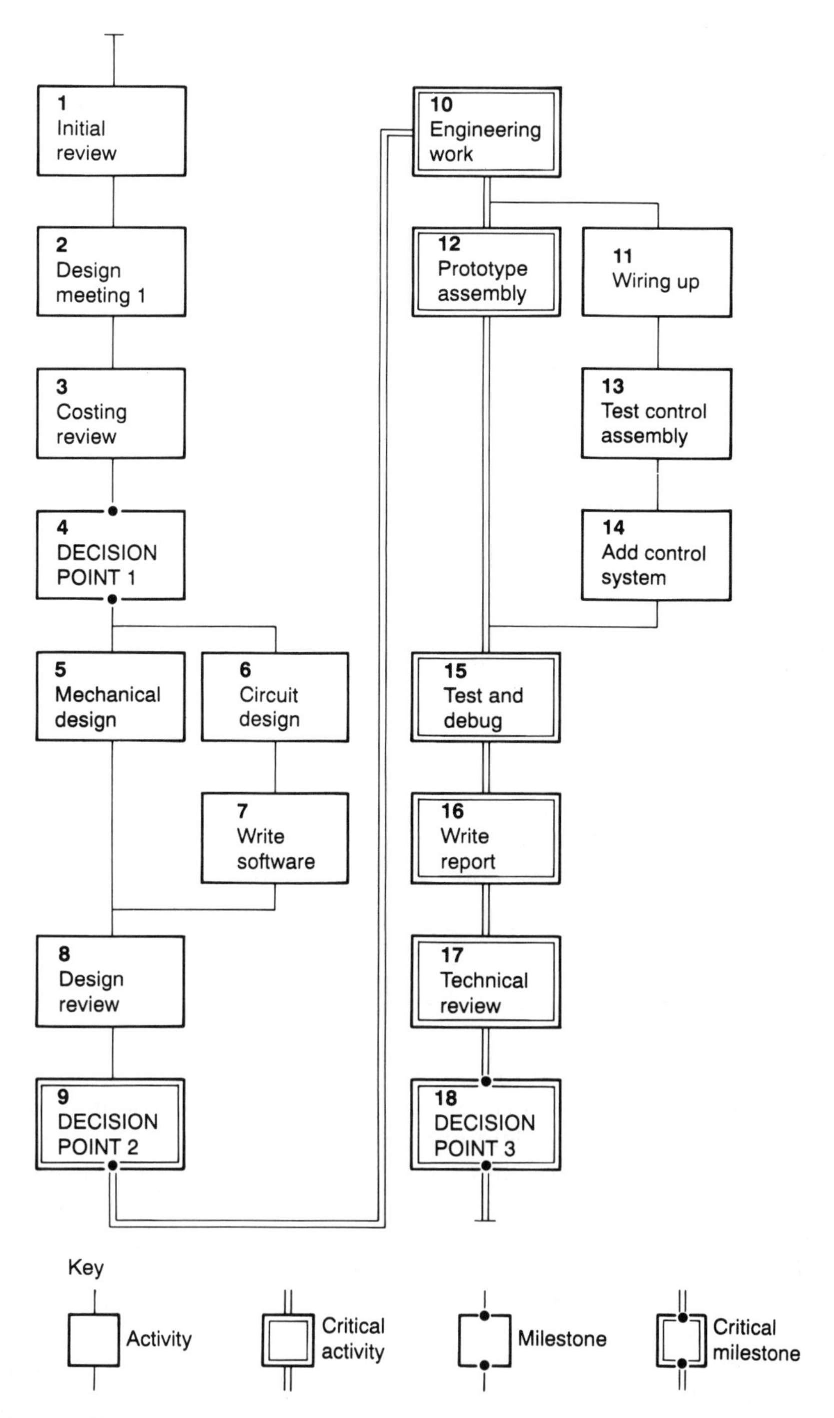

Fig. 8.1 Typical project network which shows interrelationships between activities.
(PERT chart)

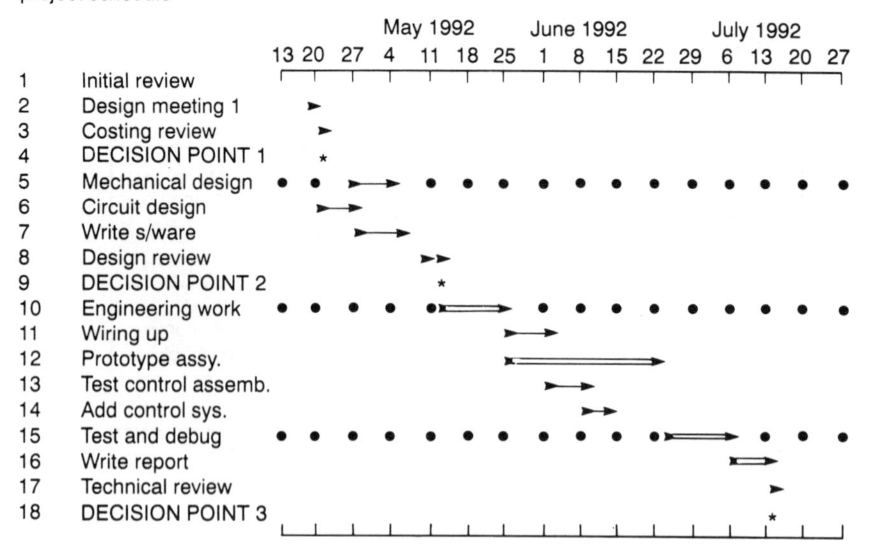

Fig. 8.2 Typical project schedule. Asterisks represent 'milestones' or decision points; double lines show the critical path
(Gantt chart)

quite raw information. It can be an excellent guide, which can be modified as the activities and their dependencies are evaluated.

Timing

In most projects time is very important. Even if there is no compulsion to complete the work to fulfil a contractual obligation, every unnecessary day costs unnecessary money in staff and resources. So the duration of each activity should be assessed, and the minimum duration of the project determined.

The time/sequence relationship is best depicted in an alternative view called a *Gantt* chart (*see* Fig. 8.2). Notice how this view shows not only the critical path (p. 188) but its duration. It also shows which activities can be programmed in at several possible places or at any convenient time, and which, in contrast, are totally dependent on the completion of a predecessor.

The timed programme thus consists of a defined sequence of activities, each of which has specific cost implications. These costs can be summed over each activity, over each phase of the project and over the whole project, thus giving a total project budget.

Review points

The project should always have a number of *review dates*. These are critical points in the programme where progress-to-date against plan can be assessed, and alternative 'fall-back' plans considered. The 'fall-back' may be as catastrophic as abandoning the project, but it could equally involve injecting extra resources to ensure that a particular deadline is met.

These alternative 'fall-back' plans should be worked out well in advance, ideally concurrently with the original, main plan (*see below*). Otherwise, unwise decisions may be taken on the spur of the moment, and panic rather than planning will rule the day.

Project Reviews are important. The project team should have had clearly defined objectives and deadlines for their activity, and should produce regular documentation to show the actual position against plan (*see* p. 195).

Review resources against timescale

Having defined the stages of the project and its timing, review the staffing and other resources in the light of the timescale. Will the staff you need be available at the right time? Where might the key technical problems occur? What limitations might there be on the availability of specialized apparatus or facilities? Which consumable materials are likely to be most difficult to get? When should they be ordered to ensure their availability at the right time?

Also, if the budget restrictions are tight – and they usually are – will there be enough cash to see the project through?

■ EVALUATE THE RISKS

All business projects have risks associated with them, and a principal reason for conducting project planning is to minimize these risks. The steps so far should have identified the areas of greatest uncertainty, and it is wise to consider what the problems of each might be, how they might be circumvented and what the financial consequences of delay might be.

Identify the risks

It is difficult to give general advice, since the problems are generally project, and usually activity, specific, but the various contingencies can range from abandoning the project, through finding additional cash to extending the completion date.

If there is a particularly critical objective, it may sometimes be possible to tackle the task in two parallel ways, so that if one is delayed or fails to work, the other acts as a back-up.

I was involved in the development of an immunoassay, in which, as is normal, the initial production of a good antibody is a very critical step: everything else depended on it. Although the team felt that a monoclonal antibody would be preferable, we set about producing antisera by both monoclonal and conventional means. If the monoclonal failed (a more tricky technique at the time), then at least we would have a conventional antibody which might be purified and used in the assay.

In the event, both methods worked – but the conventional antibody turned out to be far more appropriate to the assay!

This example above also illustrates another point. If there is a particularly critical step which can be done first, then that is the best place to do it. If that step fails and necessarily leads to the abandoning of the project, then a minimum of time and money will have been wasted.

In some instances, it may be possible to undertake a small study to assess the feasibility of a critical and expensive step which has, for some good reason, to be programmed late on in the project. It may be worth spending the money on the feasibility analysis, despite the overall additional cost it entails.

Calculate the financial consequences

The final stage is to check the finances. From the overall business planning aspect, the cost of undertaking the project must make sense against the anticipated sales of the new product. From the project management view, final project budgets must be constructed with the best information on resources needed, the cost of each and when large amounts of finance will be required.

It is also important to assess the costs of the more likely deviations from the optimum. What will the cost implications be if an extra month is needed on the most critical area? (This will not just affect the cash: it will affect the cash flow, because certain large bills may still have to be paid at the original times.) Or if we have to put an extra skilled programmer in to develop the software? Or if we have to use the more expensive method to achieve the required specification for a sub-assembly?

Such computation of 'what-if' situations is known as *sensitivity analysis*. The results are not intended to induce fear, but realism. It *is* important to know the most sensitive areas, and what the consequences might then be. In this way, particular attention is focused on the criticality of these areas, and they are more likely to receive the close attention they deserve.

This sort of multi-variate analysis, where the effect of one change in one activity cascades down into all subsequent ones, lends itself to analysis by computer. Besides the common spreadsheets, which are useful for budgeting and cost analysis, there are some good project planning programs. Some are quite inexpensive. It may therefore be worth using such a program for sensitivity analysis and general planning purposes, especially for complex projects with many interlinking parts.

However, a word of warning. Such software requires study and training before it becomes easy to use. They are by no means as intuitive as spreadsheets, and the danger is of thinking more about how to use the program than how to run the project. Rough it out with old fashioned pencil and paper first! *Then* it will be easier to put it in the computer and assess variations.

Project control

Even the best planned projects don't run themselves. They need to be monitored and controlled. Control points should be built into the plan (*see* p. 193), and these will allow analysis of variance against the plan and associated budget, and prompt identification and discussion of problems arising.

Control should be exercised right from the start, *not* from the first review point, which may be some way 'down the track': all sorts of things may have gone astray by then.

Communication is the key. The project manager should make sure that, from Day 1, each person on the team understands:

- the overall objectives. They may not need, or be able to appreciate all the details, and some may be confidential to the senior management. Nevertheless, the broad targets should be understood, and that they themselves must play a vital role in achieving them. If properly done, this initial briefing can be a powerful motivator and weld the group together as a project team.
- precisely what he or she has to do, and by when, putting it into overall context where appropriate. Ensure that they are not being set impossible deadlines, and, where they are tight, explain the urgency. Listen to the feedback, which might well indicate the presence of factors which you were not aware of. Try not to impose targets but to reach mutual agreement on them; in this way, the staff will feel some personal responsibility for meeting the deadline.

- what information they, or the person reporting on each subgroup's progress, will need to provide for each regular project report and for each review meeting. There are various standard formats for this, or it may be better to adapt one for the specific project. The information in the reports should be as quantitative as possible, with variances from activity and cost target calculated.

Quality assurance

The topic of Quality Assurance is very important to modern industry and commerce, and deserves a book to itself. This section can only indicate certain general principles as related to project work. Even attending to basic quality aspects can be of considerable value, whereas ignoring them can easily lead to disaster.

Work for some contractors such as the defence industry or in pharmaceuticals involves industry or government quality standards, so if you are undertaking a project on behalf of such an organization, there will be agreed standards and codified means of satisfying them.

It may be that your company has already instituted a Quality System, perhaps up to the standard of BS 5750. If so, any projects should be subject to the company's overall policy on quality. If you are just starting up or have not instituted a formal Quality Assurance System, then the question of quality standard in project work should be addressed.

Do not confuse Quality Assurance with Quality Control. The latter means testing the products as they come off the workbench or production line and rejecting those that do not meet a required standard. Those that are rejected are either subjected to reworking to bring them up to specification or are scrapped.

Quality Assurance means *getting it right first time.* It means checking the quality of raw materials and components, the calibration of instruments used for measurement, the standard of design and the calculations or analyses involved. It means ensuring that the staff employed are competent and trained to do the job. It means establishing a system of traceability, so that the progress of the project is monitored and documented. This kind of system is not something which is instituted overnight, and it cannot be done by delegation. In short, it implies a policy and system which is complied with and supported from the top management downward.

Theoretically (and in practice too, with fully instituted Total Quality Management), Quality Control, with all its attendant waste and expense, becomes largely unnecessary. In practical project management terms it will mean a better chance of getting the specification right, of producing a reliable product or prototype, and of doing it on time. All of which should make you and your customers happy, and enhance your company's reputation. It also saves money in the medium and longer term.

The embryonic technology-based business, with one or two managers and a few staff engaged, initially at least, in project work and prototype design and construction, are not likely to be aiming at BS 5750. Nevertheless, they want to be recognized as a quality organization, and maybe BS 5750 is something to be aspired to later, as the company grows. What can they do to aid this, to instil a 'quality ethic', at this early stage?

Many of the things are no more than applied common sense, and are things which most sensible companies would do anyway. However, it is important to do them in a 'systematized' and documented manner. The following is a checklist for a typical prototype development and construction project. It can be added to or adapted, once the general concept is appreciated:

- define the objectives, specifications and parameters of the product as quantitatively as possible;
- if the product is formed from a set of sub-assemblies, do the same for each of these;
- in drawing up sets of design, blueprints or calculations, make sure that they are checked, signed, dated and given an unambiguous number. They should then be filed. This will make the task of checking and tracing much easier.
- specify the standard of components and raw materials, and ensure that documentation certifying the specification for each consignment is received.
- document and reject components or sub-assemblies which do not meet the required specifications ('non conformity').

Most competent managers will do this sort of thing anyway, but often in a haphazard and non-systematic way. Developing a documented system overlays a valuable discipline on the project and its team. It also lays the foundation for a more formal quality system as the company grows.

Chapter 9
MARKETING

It's people that matter

■ UNDERSTAND THE MARKET

If you thumb through the various leaflets and booklets published by the banks, government agencies and so on to help small businesses, there is usually a lot on finance and administrative matters but very little on the market or marketing. This is an unusual deficiency, since the most expert management team in the world cannot run a business if the market is not there or if there is no idea on how to attack it. The reason may partly be due to the fact that the banks' major commodity and interest is money. It may also be because marketing and selling deals principally with *people,* and people and interpersonal contacts are some of the most difficult parameters to define or quantify. You cannot put customers and their desires into a spreadsheet as you can cash flow.

Nevertheless, your business – *any* business – will not be a success unless you think about satisfying the customers and how to keep on satisfying them with your products or services. It is therefore important for a book of this nature to include some general comments on marketing and selling, with later reference to technology-based products and services.

The starting point for your business is a demand for what you can supply. If you are typical, you have first thought of a product or service and have assumed people want to buy it. The average entrepreneur tends to be supply orientated, and thinks about the market demand later.

The crunch question is *why* did you decide on this particular product or service? The answer is probably because you're good at it, you could make it better than anyone else or perhaps because you've always wanted to start a business and this was the area in which you felt your special technical know-how and experience could be most successful.

All these reasons are fine, but are not enough. The other factors which must be addressed are whether there is a market and whether the people who constitute that market will buy your product at a price which will enable you to make a profit.

It should be noted at this point that selling is only one aspect of the subject of marketing. The task must begin with the Market Assessment, which is an integral part of the Business Plan (*see* Chapters 1 p. 18 and 4 p. 95). Nobody with any business sense is going to lend you money or take a stake in your enterprise unless you produce convincing evidence that there is a group of people who are likely to want your product.

■ THINK ABOUT YOUR CUSTOMERS

You don't sell to a nebulous concept called a market, you sell to *people*. The market consists of people – your potential customers. You need to ask *who* they are, *where* they are, and *why* do they buy.

Who are they?

If you are planning to launch a new product or service, *think about your customers*. What sort of people are they? What are their desires, worries, problems? Try to understand these, and see if you can project the product in such a way as to satisfy a desire or alleviate a problem.

This sort of thinking does not always come easily to the researcher or technician, but it is important for the success of a company. Successful innovation should be market-led, and the sales of the product should be customer-driven.

□ A small company in the US developed an electronic stethoscope, and felt that they could sell into a substantial, technically-literate market. They advertised the technical superiority of the product, including details of things like its enhanced frequency range.

Their initial sales were abysmal. They had ignored the simple fact that their customers, the medics, were not interested in such niceties: *their* worry was that they might make a mistake. The advertising would have produced much better results if it had focused on the reliability and certainty involved in its use. □

If you are selling to individuals, what sort of people are they? Are they professionals, manual workers, men or women, old or young?

If your company sells principally to other businesses, it is still *people* within those businesses who make the decision to buy. What sort of people are they, and what sort of people might have a key role in influencing that decision? All these factors should influence the way in which your product is projected and sold.

Where are they?

The location of customers will, of course, depend on the nature of the business and the product involved. Each product needs separate analysis: a firm accustomed to producing software for business use must totally rethink its marketing, advertising and sales campaign if it is to break into the domestic market. This may sound obvious, but large companies, which considered their names and reputations to be so well known that they would be front runners in any market, have learnt this to their cost.

Why do they buy?

'My wife thinks I bought her a fur coat to keep her warm. In fact, I bought it to keep her *quiet*!' No correspondence will be entered into on the various sociological and ethical aspects of this old music hall joke, but the point is made that we do not always buy things for the reasons we state overtly. We often deceive ourselves as well on this.

Television advertisements illustrate this very well. Besides the 'value-for-money' statements, which gives logical justification for purchase, there is usually another, much more emotional message, which implies perhaps that purchasers of the product have enhanced status, are more glamorous, higher achievers, sexier or constitute the 'in-crowd'. These messages play unashamedly on the desires of the viewer.

Of course, a simple message like 'buy my product because its better and cheaper' has an effect on the sales of many products, and if that is your main selling point, then it's not a bad start. But try using that slogan selling scent against a well-known brand of expensive perfume (even the words 'scent' and 'perfume' have connotations of status) with all the seductive imagery used in its promotion, and see who wins.

The situation is not that different if you have commercial customers. There is, perhaps, a harder-nosed financial aspect involved, but even in business, emotions such as aspirations of status are very important. 'Top people take *The Times*' is at least as powerful an incentive to buy as appealing to lower price or greater news coverage.

Frequently, there are two or more people in a company who must be simultaneously convinced, and the message for each may be different. The secretary has to be persuaded that a particular word-processor software suite will be easy to use and will so please her boss that an increased salary cannot be far away. The boss must be persuaded that it is good value for money, and that the speed of production and quality of his documents will lead to an enhanced image and greater productivity.

It is quite likely that only the secretary will study the brochures in detail, so the copy must not only first persuade her that this software is the one for her, it must also subtly feed her the arguments to use when she tackles her boss on the matter.

So: *THINK ABOUT YOUR CUSTOMERS.*

Marketing technology-based products

■ BEWARE OF TECHNOLOGICAL OBSESSION!

Marketing technology-based products really isn't very different from marketing anything else. That is why the first half of this Chapter has rarely mentioned technology. The objective in all cases is to inform the market that the product exists and what it will do, to create a desire for it (not at all the same thing) and then sell it. These are not easy tasks, and the technology can in fact make it more difficult if it is not properly handled.

Don't expect your customers to beat a path to your door just because the antibody in your analysis kit is monoclonal rather than conventional, or because your computer has the latest RISC central processor. Although technological appeal can be *part* of the marketing strategy, there are usually far more subtle, human factors which are paramount. This is one of the most important factors for the technically-minded entrepreneur to appreciate: the customers may not be so proud of or obsessed by the technological superiority of your product as you, and, if you want to sell it, you may well have to base your selling points on attributes other than the technology.

Major companies with large advertising budgets know this very well. Cars are sold to the average motorist principally on looks, comfort and reliability; the superior engineering is usually mentioned, but is there in the smaller print, as 'something which you would of course expect from our company'. It is only the marketing of the sports versions which stresses the engine performance and mechanical details.

□ The Macintosh Computer is another good example. The technology of the original 'Mac' was certainly innovative, and no doubt computer enthusiasts were impressed by the new technical thinking that had gone into it. But the computer experts were not the target market. It would have been totally counter productive to launch a major campaign based on the enhanced technology.

In fact, the converse was true. It was promoted – and very successfully – as the computer for non-experts. Even the name suggested something non-technical, and friendly. The stress was on ease of use rather than speed of number crunching. The cartoon-like icons of filing cabinets and waste bins appealed to the office workers, secretaries and home users who would normally have been totally deterred by technology. 'Mac' would be their friend and helpmate, not a piece of technology to feel apprehensive about.

The campaign was thus able to appeal to the customers' desires and, at the same time, to allay their fears: important points, which I shall return to later. I should also add that I don't use a Mac! The PC has also become very user friendly, and the advent of the Mac had an influence on this.

It is also interesting to note the comments of Brian Oakley, head of the UK's Alvey programme of collaborative research in information technology, which was undertaken between 1982 and 1988, and consumed some £200 million of public money plus £150 million from industry.

The final report on the programme, published in 1991, described many examples of first class research coupled with the lack of investment by industry to commercialize these. Oakley emphasized the need to spend more on this, especially on marketing. He specifically noted the example of the Macintosh computer, commenting that the person/computer interface developed within the Alvey programme was just as good, but no British firm would have invested the £70 million which Apple spent *on marketing alone.* □

■ EDUCATE THE MARKET

In terms of the market, technology-based products can be classified either as incremental improvements on existing products, or as revolutionary new ones. Marketing the first sort is relatively easy, since the potential customers already understand the product, and the task is to persuade them that the new product is better or cheaper, or both. To counteract this ease of market approach, the competition will be quick to respond.

Products based on entirely new principles are more difficult to market, even to users who are scientifically trained and knowledgeable. No matter how revolutionary or how much better your new product is, people get set in their ways and must be persuaded to change from something they know and are familiar with.

Sometimes they just can't believe that the product you are offering can be as good as you say it is. They therefore have to be educated to understand the new product's advantages before it can be sold to them. This can take a long time and cost a lot of money; it can prove very frustrating for the innovator, who has so clearly seen the advantages of his invention from the time of its conception.

I was closely involved in the development of some of the first commercial immunoassay kits for routine food analysis. The technique had been well demonstrated in clinical diagnosis, where it was used in literally life-and-death situations, but was entirely new to food analysts, whose training is in chemistry rather than immunology. So, although the scientific concept and the advantages of the product were not new, they were *new to the particular market we* were *addressing. This was an important lesson to learn.*

The first task was to convince them of the validity, specificity and reliability of the 'new' technology. A number of methods were used to achieve this, and it took considerable time and expense. It was necessary to organize demonstration meetings, where the new method was demonstrated, and potential users could try it 'hands on'. Experienced, well respected and independent scientists at Research Institutes evaluated the method and published their results in journals commonly read by potential customers. A number of independently run 'blind' tests involving groups of analytical laboratories were carried out and the results published.

Once this familiarity was achieved it was immediately obvious that the performance of the kits far surpassed all other methods available. However, so far as the Public Analyst is concerned, the critical test of the methods used is that they should be acceptable in a Court of Law as valid scientific evidence, and it was not until this was achieved that the products gained full acceptance.

It is pleasing to say that the assay kits are now used by Public Analysts and Food Manufacturers throughout the country and in many parts of the world, and have become the preferred method of analysis for some substances. Also, the competition has arrived, a sure mark of the commercial significance of an innovation! Thus, once the market for an entirely new product has been established, it can be profitable – but it can take a long time to get there.

The key point here is the need to educate the customer to the benefits of the product or service you provide. Not just vague, overall benefits, but the benefits to *them*. The education can be achieved by all the usual means of marketing, including demonstrations, seminars and videotapes.

■ CUSTOMER SATISFACTION IS PARAMOUNT

Your marketing effort must be *customer driven*. If you know and understand your customers, their needs, desires and concerns, then you are well on the way to selling to them.

This sounds a tall order, especially to the typical inventor, who, whilst not being the recluse he is sometimes portrayed as, is not commonly endowed with the extrovert character or interpersonal skills needed for the task. This can be overcome by employing a marketing or sales manager who has got these attributes, but even if this is not feasible initially, there are some guiding principles which may help.

Select the first customers carefully

The market for an innovative product is not homogeneous but consists of different people, some of whom are probably keen to try anything new and some of whom will be highly resistant to change. In the middle are the largest group, who are more or less persuadable but who may take time and need others to set an example for them to follow.

The people to go for first are the trend-setters, enthusiasts for novelty, the 'keenites' or early-adopters. You could waste an enormous amount of time and money trying to sell to the others. The trend-setters are the opinion leaders, and, provided they like your product or service, they will tell others. In this way you will get your first customers to do some marketing for you. (Of course, if your product is not satisfactory or they are unhappy with the service, it will be a disaster).

Once the trend-setters have by-and-large accepted and even adopted your product, a broader marketing effort will be worthwhile, to convert the more adaptable of the persuadable people in the middle of the target market. This will be more expensive, but you now have the experience of the early-adopters to aid and abet the effort.

Your product should be special

Try to make your product special in the eyes of your customers. In one sense, this should not be difficult, since it is obviously special to you. You were convinced enough of its novelty and superiority to anything else to go and start a business based on it, with all the effort and risks involved.

On the other hand, the characteristics which were special to you may not be the ones which are relevant in the view of the customers (*see* p. 201). They may perceive other factors as important, and it is advisable to think about these matters very carefully, taking advice or soundings before the market launch.

The technology will probably be secondary. Ease of use, perception of quality, price, compatibility with lifestyle or workstyle, etc., may all be more important. When the fax machine was first introduced there was little or no mention of the technology, but only of how easy it was to use and how it would improve business efficiency.

Thinking about your customers first, rather than the technology, can actually direct the subsequent technology. The thought that Sony put into its lightweight video camera/player is a good example. Their perception of the market need was not that the current machines produced a poor picture, were inflexible, difficult to use or technologically crude: they were too big and too heavy. There was a much larger market waiting for them if they could produce a camera which was just as good but which was substantially smaller and lighter.

This establishment of their customers' needs and desires for a lighter machine led to the substantial investment in design and technology to produce the product. The product could not have been produced without the technology, but the drive for it came from the market.

The *quality* of the product is also of critical importance to marketing. A technically superior product is still a disaster if it is unreliable. Any trend-setters who buy such a product will feel cheated themselves and will rapidly inform anyone who cares to listen to them. Any chance of expanding the market will then vanish.

Try to find ways of reducing reluctance to change to your product. It is quite understandable for even the most adventurous customers to have some inhibitions about changing to a new and untried product. How might these be reduced?

One way is money. A free or reduced initial offer can work just as well with a new software suite as it can with a bottle of washing-up liquid. It reduces the reluctance and risk to the customer, and indicates that you are so convinced that he will soon switch his allegiance that you are prepared to put money on it. So, consider a trial offer or a 'money-back' guarantee.

Another way, which should be played very carefully, is to raise some concern in your customer's mind about what will happen if he does not buy the new product. Will his lifestyle or business suffer in some way? Can you point to another purchaser who is already gaining advantage?

Develop a philosophy of service

Business today is immensely competitive and customers are becoming more demanding. The companies which are going to win are those which develop a philosophy of *service*, so that the customer is just as pleased with the product a year after as on the day he bought it.

We are all familiar with the company who sends a salesman who promises the earth before we buy the product, but then shows incredible reluctance to send any assistance whatsoever when it goes wrong. Have *you* ever tried to get a replacement switch for a five-year old vacuum cleaner? 'I'm sorry sir, its now an obsolete model and they don't carry any of them in stock any more.'

We can probably all continue this script for several paragraphs, but the point is, does this sort of attitude to customer service encourage the customer to return when he finally does want a totally new machine?

The same is true of back-up materials for new products. I imagine most readers will have, at some stage, bought a computer with a manual which cannot have been written with any thought at all for the customer's needs. I have one in front of me now. One page shows me, with detailed illustration, how to wire up a three-pin plug; the next immediately goes into detail on the partition tables for the hard disc. There is nothing in between. Just who is this written for? Can it possibly give customer satisfaction and assurance of service?

Things are improving now, and there is an increased perception of the need to serve the customer to maintain competitiveness. More companies are establishing Quality Assurance and Total Quality Management systems. They are offering better maintenance and advice services.

The need is to make it easy for the customer to do business with you. If you can achieve this, he will be delighted with his purchase and will come back again to buy from you.

APPENDIX
CONTACT LIST

This list represents a range of the major contact points in the UK and Ireland for those seeking information and finance for a new technology business. It is not a comprehensive list of all business support organizations. For instance, the Enterprise Agencies, many of which do sterling work, have not been included, since these do not generally concentrate their efforts on the technological areas. If they are needed, they will be easily accessible through one of the other contacts given, such as the DTI Small Firms Service or a Development Agency.

It is also better if the entrepreneur is not given a long list of contacts and sent on his way to try his luck. It is preferable to direct him to the nearest central support point, where it is hoped that the staff will themselves contact other appropriate organizations. Businessmen can find themselves spending far too much time chasing around the countryside seeking odd bits of advice and help.

The list is sorted geographically into national organizations serving the whole country, and those based in particular Counties. Within each, the addresses are coded as:

BIC: the Business and Innovation Centres of the European Community. These give a comprehensive service to clients but can be rather selective in whom they help.

DA: Development Agencies. These are able to give a range of services including finance. Their ability to deal with smaller, technology-based firms varies. They can be good initial sources of advice, and signposts to local centres such as Enterprise Agencies.

FIN: sources of financial advice and finance. The list is not a comprehensive one. Only those venture capital companies meeting certain criteria have been included (*see* Chapter 6 p. 154 and Table 6.2); only those clearing banks operating a specific investment advice service to technology firms have been listed.

Of course, you may well get just as good a service from one of the other banks – it will often depend on how well you know the manager or have managed your finances in the past that matters. This part of the Table should carry a 'Health Warning' not to approach sources of venture capital or major investment without having a sound Business Plan: read the relevant Chapters of the book before you act!

GOV: DTI and other government departments and organizations. The 'Freephone Enterprise' service of the DTI has been the starting point for many new businesses in recent years. They also run an information service on exports and the Single European Market.

PAT: organizations dealing with Patents and intellectual property. Local Patent Agents will also be listed in the 'Yellow Pages', but make sure that the one you contact is a Chartered Patent Agent.

SPARK: Science Parks. All current members of the UK Science Park Association have been included. Most have good sources of advice and finance available, but rents can be higher than elsewhere.

TEC: Training and Enterprise Councils. These are local, employer-led organizations set up by the government to support enterprise and business training. Most can provide local information and contacts, but some are still establishing themselves and their performance seems variable.

UDIL: members of the University Directors of Industrial Liaison; the contacts can facilitate access to the technological, research and other facilities of their parent university. Most polytechnics and colleges of higher education have similar units.

NATIONAL

FIN
KPMG Management Consulting
High Technology Practice
PO Box 695, 8 Salisbury Square
London EC4Y 8BB
Tel: 071 236 8000 Fax: 071 832 8888

FIN
Barclays Bank High Technology Team
PO Box 120, Longwood Close
Westwood Business Park
Coventry
Warwickshire CV4 8JN
Tel: 0203 694242 FAX: 0203 532497

FIN
Midland Enterprise, Midland Bank plc
Courtwood House
Silver Street Head
Sheffield
South Yorkshire S1 3GG
Tel: 0742 529479

FIN
National Westminster Bank Technology Unit
3rd Floor Fenchurch Exchange
8 Fenchurch Place
London EC3M 4PB
Tel: 071 374 3707

UK Science Parks Association
Aston Science Park, Love Lane
Aston Triangle
Birmingham B7 4BJ
Tel: 021 359 0981 Fax: 021 333 5852

British Venture Capital Association
3 Catharine Place
London SW1E 6DX
Tel: 071 233 5212 Fax: 071 931 0563

Business in the Community
227A City Road
London EC1V 1LX
Tel: 071 253 3716

Scottish Business in the Community
Romano House, Station Road
Corstophine
Edinburgh EH12 7AF
Tel: 031 334 9966

Confederation of British Industry (CBI)
Centrepoint
103 New Oxford Street
London WC1A 1DU
Tel: 071 379 7400

Association of British Chambers of Commerce
Sovereign House
212A Shaftesbury Avenue
London SE15 3SP
Tel: 071 240 5831

European Venture Capital Association
Keibergpark Minervastraat 6, Box 6
B-1930 Zaventem
Belgium
Tel: 01032 2 720 6010
Fax: 01032 2 725 3036

Department of Employment
Small Firms Division, Steel House
Tothill Street
London SW1H 9NF
Tel: 071 273 5000 or 0800 500200

NEDO Corporate Venturing Centre
BASE International, Midsummer House
443 Midsummer Boulevard
Milton Keynes MK9 3BN
Tel: 0908 664315

European Business & Innovation Centre Network (EBN)
Ave. de Tervueren 188A
Brussels B-1150
Belgium
Tel: 010 322 772 8900
Fax: 010 322 772 9574

PAT
Chartered Institute of Patent Agents
Staple Inn Buildings
London WC1V 7PZ

PAT
The British Library
Science Reference & Information Service
25 Southampton Buildings
London WC2A 1AY

PAT
The Patent Office Search & Advisory Service
Room 312 State House
66–71 High Holborn
London WC1R 4TP
Tel: 071 829 6000 Fax: 071 405 0292

PAT
The Technology Exchange
West Park
Silsoe
Bedfordshire MK45 4HS
Tel: 0525 60333 Fax: 0525 60664

PAT
Licensing Executives Society (GB)
Secretary, Boyden Land & Co
266 Malden Road
New Malden
Surrey KT3 6AR
Tel: 081 949 2509 Fax: 081 942 5245

ANTRIM

DA
Local Enterprise Development Unit (LEDU)
25–27 Franklin Street
Belfast
Antrim BT2 1BA
Tel: Freephone LEDU

FIN
Enterprise Equity (NI)
Bulloch House
Linenhall Street
Belfast
Antrim BT2 8PP
Tel: 0232 242500 Fax: 0232 242487

FIN
Industrial Development Board for N. Ireland
IDB House
64 Chichester Street
Belfast
Antrim BT1 4JX
Tel: 0232 233333 Fax: 0232 231328

SPARK
Antrim Technology Park
Belfast Road
Antrim BT41 1QS
Tel: 08494 61538 Fax: 08494 28075

AVON

FIN
3i plc
Bristol
Avon
Tel: 0272 277412

GOV
DTI South West
The Pithay
Bristol
Avon BS1 2PB
Tel: 0272 272666

SPARK
Bristol University Science Park
Industrial Liaison Office
3 Priory Road, Clifton
Bristol
Avon BS8 1TX
Tel: 0272 303030 Fax: 0272 732657

TEC
Avon TEC
PO Box 164, St. Lawrence House
29-31 Broad Street
Bristol
Avon BS99 7HR
Tel: 0272 277116 Fax: 0272 226664

UDIL
University of Bristol
Industrial Liaison Office
3 Priory Road, Clifton
Bristol
Avon BS8 1TX
Tel: 0272 303030 Fax: 0272 732657

UDIL
University of Bath
Technical & Commercial Marketing
Claverton Down
Bath
Avon BA2 7AY
Tel: 0225 826826 Fax: 0225 462508

BEDFORDSHIRE

SPARK
Cranfield European Technology Park
Cranfield
Bedfordshire MK43 0AL
Tel: 0234 752731 Fax: 0234 751637

TEC
Bedfordshire TEC
Woburn Court, 2 Railton Road
Woburn Road Industrial Estate
Kempston
Bedfordshire MK42 7PN
Tel: 0234 843100 Fax: 0234 843211

BERKSHIRE

FIN
Seed Capital Limited
Boston Road
Henley
Berkshire RG9 1DY
Tel: 0491 579999 Fax: 0491 579825

FIN
3i plc
Reading
Berkshire
Tel: 0734 584344

FIN
Venture Capital Report
Boston Road
Henley
Berkshire RG9 1DY
Tel: 0491 579999

SPARK
Reading University Innovation Centre
Philip Lyle Building
University of Reading
Reading
Berkshire RG6 2BX
Tel: 0734 861361 Fax: 0734 861894

TEC
Thames Valley Enterprise
6th Floor, Kings Point
120 Kings Road
Reading
Berkshire RG1 3BZ
Tel: 0734 568156 Fax: 0734 567908

BORDERS

DA
Scottish Enterprise, Borders
Wheatlands Road
Galashiels
Borders TD1 2HQ
Tel: 0896 58991 Fax: 0896 58625

BUCKINGHAMSHIRE

FIN
3i plc
Milton Keynes
Buckinghamshire
Tel: 0908 668168

TEC
Milton Keynes & North Bucks TEC
Old Market Halls, Creed Street
Wolverton
Milton Keynes
Buckinghamshire MK12 5LY
Tel: 0908 222555 Fax: 0908 222839

CAMBRIDGESHIRE

FIN
Prelude Technology Investments
280 Science Park
Milton Road
Cambridge
Cambridgeshire CB4 4WE
Tel: 0223 423132 Fax: 0223 420969

FIN
3i plc
Cambridge
Cambridgeshire
Tel: 0223 420031

SPARK
Cambridge Science Park
Bidwells
Trumpington Road
Cambridge
Cambridgeshire CB2 2LD
Tel: 0223 841841 Fax: 0223 840721

SPARK
St John's Innovation Park
The Bursary
St. John's College
Cambridge
Cambridgeshire CB2 1TP
Tel: 0223 338627 Fax: 0223 338762

TEC
Greater Peterborough TEC
Unit 4, Blenheim Court
Peppercorn Close, Lincoln Road
Peterborough
Cambridgeshire PE1 2DU
Tel: 0733 890808 Fax: 0733 890809

TEC
CAMBSTEC
Units 2–3, Trust Court, Chivers Way
The Vision Park, Histon
Cambridge
Cambridgeshire CB4 4PN
Tel: 0223 235633 Fax: 0223 235631

UDIL
Wolfson Cambridge Industrial Unit
20 Trumpington Street
Cambridge
Cambridgeshire CB2 1QA
Tel: 0223 334755 Fax: 0223 332662

CENTRAL

SPARK
Stirling University Innovation Park
Scottish Metropolitan Alpha Centre
Innovation Park
Stirling
Central FK9 4NF
Tel: 0786 70080 Fax: 0786 51030

UDIL
University of Stirling
Department of Business & Management
Stirling
Central FK9 4LA
Tel: 0786 73171 Fax: 0786 63000

CHESHIRE

BIC
Cheshire BIC
Innovation House 6 Seymour Court
Manor Park
Runcorn
Cheshire WA7 1SY
Tel: 0928 579462 Fax: 0928 579186

TEC
NorMidTEC
Spencer House, Dewhurst Road
Birchwood
Warrington
Cheshire WA3 7PP
Tel: 0925 826515 Fax: 0925 826215

TEC
South East & Cheshire TEC
PO Box 37, Middlewich Business Park
Dalton Way
Middlewich
Cheshire CW10 0HU
Tel: 0606 847009 Fax: 0606 847022

CLEVELAND

FIN
British Steel (Industry)
North of England
Cleveland House, 7 Woodlands Road
Middlesborough
Cleveland TS1 3BH
Tel: 0642 244633 Fax: 0642 244446

FIN
3i plc
Hull
Cleveland
Tel: 0482 27066

SPARK
Newlands Centre
Direct, University of Hull
Cottingham Road
Hull
Cleveland HU6 7RX
Tel: 0482 465139 Fax: 0482 466666

SPARK
Belasis Hall Technology Park
Belasis Court
Billingham
Cleveland TS23 4AE
Tel: 0642 370301 Fax: 0642 370288

TEC
Teeside TEC
Corporation House
73 Albert Road
Middlesbrough
Cleveland TS1 2RU
Tel: 0642 231023 Fax: 0642 232480

CLWYD

BIC
Newtech Innovation Centre
Newtech Science Park
Deeside Industrial Park
Clwyd CH5 2NT
Tel: 0244 289881 Fax: 0244 280002

DA
Welsh Development Agency
Wrexham Industrial Estate
Wrexham
Clwyd LL13 9UF
Tel: 0978 661011 Fax: 0978 661007

SPARK
Newtech Science Park
Deeside Industrial Park
Clwyd CH5 2NT
Tel: 0244 289881 Fax: 0244 280002

SPARK
Wrexham Technology Park
Croesnewydd Hall
Wrexham
Clwyd LL13 7PY
Tel: 0978 290694 Fax: 0978 290705

TEC
North East Wales TEC
Wynnstay Block, Hightown Barracks
Kingsmill Road
Wrexham
Clwyd LL13 8BH
Tel: 0978 290049 Fax: 0978 290061

CO. DOWN

DA
Local Enterprise Development Unit
(LEDU)
6–7 The Mall
Newry
Co. Down BT34 1BX
Tel: Freephone LEDU

CUMBRIA

SPARK
Westlakes Science & Technology
Park
Cumberland House
Lowther Street
Whitehaven
Cumbria CA28 7AH
Tel: 0946 67089 Fax: 0946 64282

TEC
Cumbria TEC
Venture House, Regents Court
Guard Street
Workington
Cumbria CA14 4EW
Tel: 0900 66991 Fax: 0900 604027

DERBYSHIRE

FIN
Derbyshire Enterprise Board
95 Sheffield Road
Chesterfield
Derbyshire S41 7JH
Tel: 0246 207390 Fax: 0246 221080

TEC
South Derbyshire TEC
St. Peters House
Gower Street
Derby
Derbyshire DE1 5SB
Tel: 0332 290550 Fax: 0332 292188

TEC
North Derbyshire TEC
Block C, St. Marys Court
St. Marys Gate
Chesterfield
Derbyshire S41 7TD
Tel: 0246 551158 Fax: 0246 238489

DERRY

BIC
NORIBIC
Springrowth House, Ballinska Road
Springtown Industrial Estate
Derry
BT48 0NA
Tel: 0504 264242 Fax: 0504 269025

DA
Local Enterprise Development Unit
(LEDU)
13 Shipquay Street
Londonderry
Derry BT48 6DJ
Tel: Freephone LEDU

UDIL
University of Ulster
Research & Consultancy Services
Coleraine
Derry BT52 1SA
Tel: 0265 44141 Fax: 0265 40905

DEVON AND CORNWALL

FIN
3i plc
Exeter
Devon
Tel: 0392 438834

UDIL
University of Exeter
Exeter Enterprises
Hailey Wing
Exeter
Devon EX4 4QJ
Tel: 0392 214085 Fax: 0392 264375

TEC
Devon & Cornwall TEC
Foliot House, Brooklands
Budshill Road, Crownhill
Plymouth
Devon PL6 5XR
Tel: 0752 767929 Fax: 0752 770925

DORSET

SPARK
Winfrith Technology Centre
Winfrith
Dorchester
Dorset DT2 8DH
Tel: 0305 251888 Fax: 0305 202094

TEC
Dorset TEC
25 Oxford Road
Bournemouth
Dorset BH8 8EY
Tel: 0202 299284 Fax: 0202 299457

DURHAM

SPARK
Durham Mountjoy Research Centre
Unit 1A
Mountjoy Research Centre
Durham DH1 3SW
Tel: 091 384 4173 Fax: 091 384 0974

TEC
County Durham TEC
Valley Street North
Darlington
Durham DL1 1TJ
Tel: 0325 351166 Fax: 0325 381362

UDIL
University of Durham
Industrial Research Laboratories
South Road
Durham
Durham DH1 3LE
Tel: 091 374 2000 Fax: 091 374 2581

DYFED

DA
Welsh Development Agency
Cillefwr Industrial Estate
Johnstown
Carmarthen
Dyfed SA31 3RB
Tel: 0267 235642

SPARK
Aberystwyth Science Park
Development Board for Rural Wales
Ladywell House
Newtown
Dyfed SY16 1JB
Tel: 0686 626965 Fax: 0686 627889

UDIL
University College of Wales
Industrial Liaison Office
Penglais
Aberystwyth
Dyfed SY23 3DD
Tel: 0970 622382 Fax: 0970 617172

E. SUSSEX

UDIL
University of Sussex
Research & Industry Support Unit
Falmer
Brighton
East Sussex BN1 9RH
Tel: 0273 606755 Fax: 0273 678335

EIRE

UDIL
Trinity College Dublin
Innovation Services
O'Reilly Institute
Dublin 2
Eire
Tel: 772941 Fax: 798039

UDIL
University College Galway
Industrial Liaison Office
Galway
Eire
Tel: 91 24411 Fax: 91 25200

UDIL
University College Dublin
Industrial Liaison Office
West Theatre, Trinity College
Dublin
Eire
Tel: 276871 Fax: 275948

BIC
Galway Business Innovation Centre
Hynes Building
St. Augustine Street
Galway
Ireland
Tel: 353 91 67974 Fax: 353 91 61963

BIC
Limerick Innovation Centre
Enterprise House, Plassey
Technology Park
Castletroy
Limerick
Ireland
Tel: 353 61 338177 Fax: 353 61 338065

BIC
Dublin Business Innovation Centre
The Tower, IDA Enterprise Centre
Pearse Street
Dublin 2
Ireland
Tel: 353 1 713111 Fax: 353 1 713330

BIC
Southwest Business & Technology
Centre
IDA Enterprise Centre
North Mall
Cork City
Ireland
Tel: 353 21 397711 Fax: 353 21 395393

ESSEX

TEC
Essex TEC
Redwing House, Hedgerows
Business Park
Colchester Road
Chelmsford
Essex
Tel: 0245 450123

UDIL
University of Essex
Industrial Liaison Office
Wivenhoe Park
Colchester
Essex CO4 3SQ
Tel: 0206 873333 Fax: 0206 873598

FIFE

UDIL
University of St. Andrews
Centre for External Services
66 North Street
St. Andrews
Fife KY16 9AH
Tel: 0334 76161 Fax: 0334 75892

GLOUCESTERSHIRE

TEC
Gloucestershire TEC
Conway House
33–35 Worcester Street
Gloucester
Gloucestershire GL1 3AJ
Tel: 0452 524488 Fax: 0452 307144

GRAMPIAN

DA
Grampian Enterprise
27 Albyn Place
Aberdeen
Grampian AB1 1YL
Tel: 0224 211500 Fax: 0244 213417

FIN
3i plc
Aberdeen
Grampian
Tel: 0224 638666

SPARK
Aberdeen Technology Transfer
25 Albert Terrace
Aberdeen
Grampian AB1 1XY
Tel: 0224 641953 Fax: 0224 641953

UDIL
University of Aberdeen
Industrial Liaison Office, Industry House
48 College Bounds
Aberdeen
Grampian AB9 2TT
Tel: 0224 272484 Fax: 0224 487658

GREATER MANCHESTER

BIC
GM-BIC
CCE Business Centre, Windmill Lane
Denton
Tameside
Greater Manchester M34 3OS
Tel: 061 337 8648 Fax: 061 337 8651

FIN
British Technology Group
Enterprise House, Manchester Science Park
Lloyd Street North
Manchester
Greater Manchester M15 4EN
Tel: 061 226 2811

FIN
3i plc
Manchester
Greater Manchester
Tel: 061 833 9511

GOV
DTI North West (Manchester)
75 Mosley Street
Manchester
Greater Manchester M2 3HR
Tel: 061 838 5000

SPARK
Manchester Science Park
Enterprise House
Lloyd Street North
Manchester
Greater Manchester M15 4EN
Tel: 061 226 1000 Fax: 061 226 1001

SPARK
Bolton Technology Exchange
Unit 4
Queensbrook
Bolton
Greater Manchester BL1 4AY
Tel: 0204 28851 Fax: 0204 399074

TEC
Stockport/High Peak TEC
1 St. Peters Square
Stockport
Greater Manchester SK1 1NN
Tel: 061 477 8830 Fax: 061 480 7243

TEC
METROTEC (Wigan)
Buckingham Row
Northway
Wigan
Greater Manchester WN1 1XX
Tel: 0942 36312 Fax: 0942 821410

TEC
Rochdale TEC
St. James Place
160–162 Yorkshire Street
Rochdale
Greater Manchester OL16 2DL
Tel: 0706 44909 Fax: 0706 44979

TEC
Oldham TEC
Block D, 3rd Floor
Brunswick Square, Union Street
Oldham
Greater Manchester OL1 1DE
Tel: 061 620 0006 Fax: 061 620 0030

TEC
Bolton/Bury TEC
Bayley House
St. Georges Square
Bolton
Greater Manchester BL1 2HB
Tel: 0204 397350 Fax: 0204 363212

UDIL
University of Salford
CAMPUS
43 The Crescent
Salford
Greater Manchester M5 4WT
Tel: 061 743 1727 Fax: 061 745 7808

UDIL
UMIST
UMIST Ventures
PO Box 88
Manchester
Greater Manchester M60 1QD
Tel: 061 200 3055 Fax: 061 228 7040

UDIL
University of Manchester
Industrial Liaison Centre
Manchester Science Park
Manchester
Greater Manchester M15 4EN
Tel: 061 226 5216 Fax: 061 226 4766

UDIL
University of Manchester
Marinetech North West
Coupland III Building
Manchester
Greater Manchester M13 9PL
Tel: 061 273 3278 Fax: 061 273 8788

TEC
Manchester TEC
Boulton House
17–21 Chorlton Street
Manchester M1 3HY
Tel: 061 236 7222 Fax: 061 236 8878

GWENT

DA
Welsh Development Agency
Caradog House
Cleppa Park
Newport
Gwent NP1 9UG
Tel: 0633 815555

FIN
British Steel (Industry)
Wales/West Midlands
Clarence House, Clarence Place
Newport
Gwent NP9 7AA
Tel: 0633 244001 Fax: 0633 246278

TEC
Gwent TEC
Glyndwr House, Unit B2
Cleppa Park
Newport
Gwent NP9 1YE

GWYNEDD

DA
Welsh Development Agency
Llys-y-Bont
Parc Menai
Bangor
Gwynedd LL57 4BN
Tel: 0248 670076 Fax: 0248 671197

SPARK
Menai Technology Enterprise Centre
Ffordd Deiniol
Bangor
Gwynedd LL57 2UP
Tel: 0286 672255 Fax: 0248 352497

TEC
North West Wales TEC
Llys Brittania
Parc Menai
Bangor
Gwynedd LL57
Tel: 0248 671444 Fax: 0248 670889

UDIL
University College of North Wales
Industrial & Commercial Services
Bangor
Gwynedd LL57 2UW
Tel: 0248 351151 Fax: 0248 352497

HAMPSHIRE

FIN
3i plc
Southampton
Hampshire
Tel: 0703 632044

SPARK
Chilworth Research Centre
The Cottage
Chilworth
Southampton
Hampshire SO1 7JF
Tel: 0703 767420 Fax: 0703 766190

TEC
Hampshire TEC
25 Thackeray Mall
Fareham
Hampshire PO16 0PQ
Tel: 0329 285921 Fax: 0329 237733

UDIL
University of Southampton
Office of Industrial Affairs
Highfield
Southampton
Hampshire SO9 5NH
Tel: 0703 592296 Fax: 0703 559308

HEREFORD & WORCESTER

TEC
HAWTEC
Haswell House
St. Nicholas Street
Worcester
Worcester WR1 1UW
Tel: 0905 723200 Fax: 0905 613338

TEC
Central England TEC
The Oaks
Clewes Road
Redditch
Hereford & Worcester B98 7ST
Tel: 0527 545415 Fax: 0527 543032

HERTFORDSHIRE

TEC
Hertfordshire TEC
New Barnes Mill
Cotton Mill Lane
St. Albans
Hertfordshire AL1 2HA
Tel: 0727 52313 Fax: 0727 41449

HIGHLAND

DA
Highlands & Islands Development Board
Bridge House
27 Bank Street
Inverness
Highland IV1 1QR
Tel: 0463 234171

HUMBERSIDE

TEC
Humberside TEC
The Maltings, Silvester Square
Silvester Street
Hull
Humberside HU1 3HL
Tel: 0482 226491 Fax: 0482 213206

UDIL
University of Hull
Industrial & Commercial Development Agency
Cottingham Road
Hull
Humberside HU6 7RX
Tel: 0482 465510 Fax: 0482 465936

ISLE OF WIGHT

TEC
Isle of Wight TEC
Mill Court
Furrlongs
Newport
Isle of Wight PO30 2AA
Tel: 0983 822818 Fax: 0983 527063

KENT

FIN
3i plc
Maidstone
Kent
Tel: 0273 23164

TEC
Kent TEC
5th Floor, Mountbatten House
28 Military Road
Chatham
Kent ME4 4JE
Tel: 0634 844411 Fax: 0634 830991

UDIL
University of Kent
Kent Scientific & Industrial Products
Physics Laboratory
Canterbury
Kent CT2 7NR
Tel: 0227 475000 Fax: 0227 459025

LANARKSHIRE

DA
Lanarkshire Development Agency
Rowantree House
Newhouse Industrial Estate
Motherwell
Lanarkshire ML1 5RX
Tel: 0698 732637 Fax: 0698 733571

LANCASHIRE

BIC
Lancashire BIC
Suites 302–304, Dalsyfield Business Centre
Appleby Street
Blackburn
Lancashire BB1 3BL
Tel: 0254 692692 Fax: 0254 692290

FIN
Lancashire Enterprises
Enterprise House, 17 Ribblesdale Place
Winckley Square
Preston
Lancashire PR1 3NA
Tel: 0772 203020 Fax: 0772 204129

TEC
LAWTEC
96 Lancaster Road
Preston
Lancashire PR1 1HE
Tel: 0772 200035 Fax: 0772 54801

TEC
ELTEC
Suite 507, Glenfield Park
Site 2, Blakewater Road
Blackburn
Lancashire BB1 5QH
Tel: 0254 61471 Fax: 0254 682852

UDIL
University of Lancaster
Commercial & Industrial Development Bureau
Lancaster
Lancashire LA1 4YW
Tel: 0524 65201 Fax: 0524 843087

LEICESTERSHIRE

FIN
3i plc
Leicester
Leicestershire
Tel: 0533 555110

SPARK
Loughborough Technology Centre
Leicestershire County Council,
County Hall
Glenfield
Leicester
Leicestershire LE3 8RJ
Tel: 0533 657012 Fax: 0533 314186

TEC
Leicestershire TEC
1st Floor, Rutland Centre
Halford Court
Leicester
Leicestershire LE1 1TQ
Tel: 0553 538616 Fax: 0553 515226

UDIL
Loughborough University of Technology
Loughborough Consultants
Loughborough
Leicestershire LE11 3TF
Tel: 0509 222597 Fax: 0509 231983

UDIL
University of Leicester
Centre for Enterprise
Leicester
Leicestershire LE1 7RH
Tel: 0533 522408 Fax: 0533 522200

LINCOLNSHIRE

TEC
Lincolnshire TEC
5th Floor, Wigford House
Brayford Wharf East
Lincoln
Lincolnshire LN5 7AY
Tel: 0522 532266 Fax: 0522 510534

LONDON

FIN
3i plc
London
Tel: 071 928 7822

FIN
Korda and Company
5th Floor, 18–20 Farringdon Lane
London EC1R 3AU
Tel: 071 253 5882 Fax: 071 251 4837

FIN
Alta Berkeley Associates
9/10 Savile Row
London W1X 1AF
Tel: 071 734 4884 Fax: 071 734 6711

FIN
3i plc (Headquarters)
91 Waterloo Road
London SE1 8XP
Tel: 071 928 3131 Fax: 071 928 0058

FIN
County NatWest Ventures
135 Bishopsgate
London EC2M 3UR
Tel: 071 375 5000

FIN
British Technology Group
101 Newington Causeway
London SE1 6BU
Tel: 071 403 6666 Fax: 071 403 7586

FIN
Alan Patricof Associates
24 Upper Brook Street
London W1Y 1PD
Tel: 071 872 6300 Fax: 071 629 9035

FIN
Thompson Clive & Partners
24 Old Bond Street
London W1X 3DA
Tel: 071 491 4809 Fax: 071 493 9172

FIN
British Steel (Industry)
London Office
Canterbury House,
2–6 Sydenham Road
Croydon CR9 2LJ
Tel: 081 686 2311 Fax: 081 680 8616

FIN
Greater London Enterprise
63–67 Newington Causeway
London SE1 6BD
Tel: 071 403 0300 Fax: 071 403 1742

FIN
UCL Ventures
University College London
Gower Street
London WC1E 6BT
Tel: 071 272 8051 Fax: 071 380 7220

FIN
Barclays Venture Capital Unit
Clerkenwell House
67 Clerkenwell Road
London EC1R 5BH
Tel: 071 242 4900 Fax: 071 242 2048

FIN
FutureStart
20–21 Tooks Court
Cursitor Street
London EC4A 1LB
Tel: 071 242 9900 Fax: 071 405 2863

FIN
Barnes Thompson Management Services
120 Wigmore Street
London W1H 9FD
Tel: 071 487 3870 Fax: 071 487 3860

FIN
Local Investment Networking Company (LINC)
4 Snow Hill
London
London EC1A 2BS
Tel: 071 236 3000

FIN
Hambros Advanced Technology Trust
20–21 Tooks Court
Cursitor Street
London EC4A 1LB
Tel: 071 242 9900 Fax: 071 405 2863

GOV
Department of Trade & Industry
Room 230, Kingsgate House
London SW1E 6SW
Tel: 071 215 5000

SPARK
UCL Ventures
University College London
Gower Street
London WC1E 6BT
Tel: 071 272 8051 Fax: 071 380 7220

SPARK
South Bank Technopark
90 London Road
London SE1 6LN
Tel: 071 928 2900 Fax: 071 928 0589

TEC
London East TEC
Cityside House
40 Adler Street
London E1 1EE
Tel: 071 377 1866 Fax: 071 377 8003

TEC
CILNTEC
c/o Employment Dept. TEED
1st Floor, 236 Grays Inn Road
London WC1X 8HL
Tel: 071 837 3311 Fax: 071 837 0629

TEC
South Thames TEC
200 Great Dover Street
London SE1 4YB
Tel: 071 403 1990 Fax: 071 378 1590

TEC
West London TEC
Employment Dept. TEED, W. London Area
Lyric House, 149 Hammersmith Road
London W14 0QT
Tel: 071 602 7227 Fax: 071 603 7933

TEC
Central London TEC
12 Grosvenor Crescent
London SW1X 7EE
Tel: 071 411 3500 Fax: 071 411 3555

TEC
SOLOTEC (South London)
Lancaster House
7 Elmfield Road, Bromley
Kent
London BR1 1LT
Tel: 081 313 9232 Fax: 081 313 9245

TEC
North London TEC
Employment Dept. TEED,
N. London Area
19–29 Woburn Place
London WC1 0LU
Tel: 071 837 1288 Fax: 071 837 6518

TEC
North West London TEC
Employment Dept. TEED,
N. London Area
19–29 Woburn Place
London WC1 0LU
Tel: 071 837 1288 Fax: 071 837 6518

UDIL
University College London
UCL
5 Gower Street
London WC1E 6HA
Tel: 071 636 7668 Fax: 071 637 7921

UDIL
London School of Economics
LSE Research
Houghton Street
London WC2A 2AE
Tel: 071 831 4262 Fax: 071 242 0392

UDIL
Imperial College
Industrial Liaison Office
Room 539 Sherfield Building
London SW7 2AZ
Tel: 071 589 5111 Fax: 071 589 3553

UDIL
King's College, London
KCL Research Enterprises
Campden Hill Road
London W8 7AH
Tel: 071 937 8314 Fax: 071 937 7783

UDIL
City University
City Consultancy Services
Northampton Square
London EC1V 0HB
Tel: 071 608 4399 Fax: 071 608 2745

LOTHIAN

DA
Lothian & Edinburgh Enterprise
Apex House
99 Haymarket Terrace
Edinburgh
Lothian ED12 5HD
Tel: 031 331 4000 Fax: 031 313 4231

FIN
3i plc
Edinburgh
Lothian
Tel: 031 226 7092

FIN
Lothian Enterprises
21 Ainslie Place
Edinburgh
Lothian EH3 6AJ
Tel: 031 220 2100 Fax: 031 225 2658

FIN
British Technology Group
23 Chester Street
Edinburgh
Lothian EH3 7ET
Tel: 031 220 2860

SPARK
Heriot-Watt Research Park
Heriot-Watt University
Riccarton
Edinburgh
Lothian EH14 4AP
Tel: 031 449 7070 Fax: 031 449 7076

UDIL
Heriot-Watt University
UNILINK
Riccarton
Edinburgh
Lothian EH14 4AS
Tel: 031 449 5111 Fax: 031 451 3129

UDIL
University of Edinburgh
UnivEd Technologies
16 Buccleuch Place
Edinburgh
Lothian EH8 9LN
Tel: 031 667 1011 Fax: 031 662 4061

MERSEYSIDE

FIN
3i plc
Liverpool
Merseyside
Tel: 051 236 2944

GOV
DTI North West (Liverpool)
Greame House
Derby Square
Liverpool
Merseyside L2 7UP
Tel: 051 227 4111

SPARK
Merseyside Innovation Centre
121 Mount Pleasant
Liverpool
Merseyside L3 5TF
Tel: 051 708 0123 Fax: 051 707 0230

TEC
QUALITEC (St. Helens)
PO Box 113
Canal Street
St. Helens
Merseyside
WA10 3LN
Tel: 0744 696300 Fax: 0744 696320

TEC
Merseyside TEC
3rd Floor, Tithebarn House
Tithebarn Street
Liverpool
Merseyside L2 2NZ
Tel: 051 236 0026 Fax: 051 236 4013

TEC
CEWTEC
Block 4, Woodside Business Park
Birkenhead
Wirral
Merseyside L41 1EH
Tel: 051 650 0555 Fax: 051 650 0777

UDIL
University of Liverpool
Research, Support & Industrial Liaison
Senate House, PO Box 147
Liverpool
Merseyside L69 3BX
Tel: 051 794 2080 Fax: 051 708 6502

MID GLAMORGAN

DA
Welsh Development Agency
QED Centre
Treforest Industrial Estate
Pontypridd
Mid Glamorgan CF37 4YR
Tel: 0443 841408

DA
Welsh Development Agency
Business Centre
Business Triangle Park
Pentrebach
Merthyr Tydfil
Mid Glamorgan CF48 4YB
Tel: 0685 722177

TEC
Mid Glamorgan TEC
Unit 17–20 Centre Court,
Main Avenue
Treforest Industrial Estate
Pontypridd
Mid Glamorgan CF37 5YL
Tel: 0443 841594 Fax: 0443 841578

MIDDLESEX

FIN
Venture Founders
West Court, Salamander Quay
Harefield
Uxbridge
Middlesex UB9 6N
Tel: 0895 824015 Fax: 0895 823099

FIN
3i plc
Watford
Middlesex
Tel: 0923 33232

SPARK
Brunel Science Park
Brunel University
Cleveland Road
Uxbridge
Middlesex UB8 3PH
Tel: 0895 272192 Fax: 0895 256581

UDIL
Brunel University
Research Services Bureau
Uxbridge
Middlesex UB8 3PH
Tel: 0895 39234 Fax: 0895 32806

NORFOLK

TEC
Norfolk and Waveney TEC
Partnership House, Unit 10
Norwich Business Park
Norwich
Norfolk NR4 6DJ
Tel: 0603 763812 Fax: 0603 763813

UDIL
University of East Anglia
The Registry
Norwich
Norfolk NR4 7TJ
Tel: 0603 56161 Fax: 0603 58553

NORTH YORKSHIRE

SPARK
York Science Park
Bursar, University of York
Heslington
York
North Yorkshire YO1 5DD
Tel: 0904 59861 Fax: 0904 433433

TEC
North Yorkshire TEC
TEC House, 7 Pioneer Business Park
Amy Johnson Way, Clifton Moorgate
York
North Yorkshire YO3 8TN
Tel: 0904 691939 Fax: 0904 690411

UDIL
University of York
Industrial Development Office
York
North Yorkshire YO1 5DD
Tel: 0904 433245 Fax: 0904 432917

NORTHAMPTONSHIRE

TEC
Northamptonshire TEC
Royal Pavilion,
Summer House Pavilion
Moulton Park Ind. Estate
Northampton
Northamptonshire NN3 1WD
Tel: 0604 671200 Fax: 0604 670361

NORTHUMBERLAND

TEC
Northumberland TEC
Suite 2, Craster Court
Manor Walk Shopping Centre
Cramlingham
Northumberland NE23 6XX
Tel: 0670 713303 Fax: 0670 713323

NOTTINGHAMSHIRE

BIC
Nottingham BIC
13 Faraday Buildings,
Highfields Science Park
University Boulevard
Nottingham
Nottinghamshire NG7 2QP
Tel: 0602 436643 Fax: 0602 220718

FIN
3i plc
Nottinghamshire
Tel: 0602 412766

GOV
DTI East Midlands
Severns House
20 Middle Pavement
Nottingham
Nottinghamshire

SPARK
Highfields Science Park
17 Heathcoat Building
Highfields Science Park
Nottingham
Nottinghamshire NG7 2QJ
Tel: 0602 484848 Fax: 0602 420825

TEC
North Nottinghamshire TEC
1st Floor, Block C, Edinstowe House
High Street, Edinstowe
Mansfield
Nottinghamshire NG21 9PR
Tel: 0623 824624 Fax: 0623 824070

TEC
Greater Nottingham TEC
Lambert House
Talbot Street
Nottingham
Nottinghamshire NG1 5GL
Tel: 0602 413313 Fax: 0602 484589

UDIL
University of Nottingham
Industrial & Business Liaison Office
University Park
Nottingham
Nottinghamshire NG7 2RD
Tel: 0602 484848 Fax: 0602 420825

OMAGH

DA
Local Enterprise Development Unit (LEDU)
15 High Street
Omagh BT78 1BA
Tel: Freephone LEDU

OXFORDSHIRE

FIN
Oxford Seedcorn Capital
213 Woodstock Road
Oxford
Oxfordshire OX2 7AD
Tel: 0865 53535 Fax: 0865 512976

SPARK
Harwell Research Campus
Abingdon
Oxfordshire OX11 0RA
Tel: 0235 821111 Fax: 0235 432201

SPARK
Oxford Science Park
The Magdalen Centre
Robert Robinson Avenue
Oxford
Oxfordshire OX4 4GA
Tel: 0865 784000 Fax: 0865 784004

TEC
Heart of England TEC
26/27 The Quadrant
Abingdon Science Park
Abingdon
Oxfordshire OX14 3YS
Tel: 0235 553249 Fax: 0235 555706

POWYS

DA
Mid Wales Development
Ladywell House
Newtown
Powys SY16 1JB
Tel: 0686 626965

TEC
Powys TEC
1st Floor, St. Davids House
Newtown
Powys SY16 1RB
Tel: 0686 622494 Fax: 0686 622716

SHROPSHIRE

TEC
Shropshire TEC
2nd Floor, Hazledene House
Central Square
Telford
Shropshire TF3 4JJ
Tel: 0952 291471 Fax: 0952 291437

SOMERSET

TEC
Somerset TEC
Crescent House
3–7 The Mount
Taunton
Somerset TA1 3TT
Tel: 0823 259121 Fax: 0823 256174

SOUTH GLAMORGAN

BIC
SE Wales BIC
Cardiff Business Technology Centre
Senghennydd Road, Cathays Park
Cardiff
South Glamorgan CF4 2AY
Tel: 0222 372311 Fax: 0222 373436

DA
Welsh Development Agency
Pearl House
Greyfriars Road
Cardiff
South Glamorgan CF1 3XX
Tel: 0222 222666 Fax: 0222 223243

FIN
3i plc
Cardiff
South Glamorgan
Tel: 0222 394541

GOV
Welsh Office Industry Department
New Crown Building
Cathays Park
Cardiff
South Glamorgan CF1 3NQ
Tel: 0222 825111

SPARK
Cardiff Technology Centre
Senghennydd Road
Cathays Park
Cardiff
South Glamorgan CF2 4AY
Tel: 0222 372311 Fax: 0222 373436

TEC
South Glamorgan TEC
5th Floor, Phase 1 Building
Ty Glas Road, Llanishen
Cardiff
South Glamorgan CF4 5PJ
Tel: 0222 755744 Fax: 0222 764459

UDIL
University College of Wales Cardiff
University Industry Centre
57 Park Place
Cardiff
South Glamorgan CF1 3AT
Tel: 0222 874837 Fax: 0222 874189

SOUTH YORKSHIRE

BIC
Barnsley BIC
Innovation Way
Barnsley
South Yorkshire S75 1JL
Tel: 0226 249590 Fax: 0226 249625

FIN
3i plc
Sheffield
South Yorkshire
Tel: 0742 680571

FIN
Doncaster Enterprise Agency
19/21 Hallgate
Doncaster
South Yorkshire DN1 3NA
Tel: 0302 340320 Fax: 0302 344740

FIN
British Steel (Industry)
Yorkshire/Humberside
Bridge House, Bridge Street
Sheffield
South Yorkshire S3 8NS
Tel: 0742 700933 Fax: 0742 701390

SPARK
Sheffield Science Park
Arundel Street
Sheffield
South Yorkshire S1 2NT
Tel: 0742 724140 Fax: 0742 720379

TEC
Sheffield TEC
1st Floor, Don House
Pennine Centre, 20–22 Hawley St
Sheffield
South Yorkshire S1 3GA
Tel: 0742 701911 Fax: 0742 752634

TEC
Rotherham TEC
Moorgate House
Moorgate Road
Rotherham
South Yorkshire S60 2EN
Tel: 0709 830511 Fax: 0709 362519

TEC
Barnsley/Doncaster TEC
Conference Centre
Eldon Street
Barnsley
South Yorkshire S70 2JL
Tel: 0226 248088 Fax: 0226 291625

UDIL
University of Sheffield
Commercial & Industrial
Development Bureau
Western Bank
Sheffield
South Yorkshire S10 2TN
Tel: 0742 768555 Fax: 0742 725004

STAFFORDSHIRE

TEC
Staffordshire TEC
Moorlands House, 24 Trinity Street
Hanley
Stoke
Staffordshire ST1 5LN
Tel: 0782 202733 Fax: 0782 286215

UDIL
University of Keele
Research Development & Business
Affairs
Keele
Staffordshire ST5 5BG
Tel: 0782 621111 Fax: 0782 613847

STRATHCLYDE

BIC
Strathclyde Innovation
Unit A1, Building 1, Templeton
Business Centre
62 Templeton Street
Glasgow
Strathclyde G40 1DA
Tel: 041 554 5995 Fax: 041 556 6320

DA
The Scottish Business Shop (Scottish
Enterprise)
21 Bothwell Street
Glasgow
Strathclyde G2 6NR
Tel: 0800 222 999

DA
Scottish Enterprise
120 Bothwell Street
Glasgow
Strathclyde G2 7JP
Tel: 041 248 2700 Fax: 041 221 3217

FIN
3i plc
Glasgow
Strathclyde
Tel: 041 248 4456

FIN
Scottish Enterprise
120 Bothwell Street
Glasgow
Strathclyde G2 7JP
Tel: 041 248 2700 Fax: 041 204 3648

FIN
Venture Capital Club, Strathclyde Innovation
Building 1,
Templeton Business Centre
62 Templeton Street
Glasgow
Strathclyde G40 1DA
Tel: 041 554 5995 Fax: 041 556 6320

FIN
British Steel (Industry) Scotland
41 Oswald Street
Glasgow
Strathclyde G1 1PA
Tel: 041 221 3372 Fax: 041 221 0628

GOV
Industry Department for Scotland
Alhambra House
45 Waterloo Street
Glasgow
Strathclyde G2 6AT
Tel: 041 248 2855

SPARK
West of Scotland Science Park
6.06 Kelvin Campus
Glasgow
Strathclyde G20 0SP
Tel: 041 946 7161 Fax: 041 945 1591

UDIL
University of Glasgow
Industrial & Commercial Development Service
2 The Square
Glasgow
Strathclyde G12 8QQ
Tel: 041 330 5199 Fax: 041 330 5643

UDIL
University of Strathclyde
Research & Development Services
50 George Street
Glasgow
Strathclyde G1 1BA
Tel: 041 552 4400 Fax: 041 552 0775

SUFFOLK

TEC
Suffolk TEC
2nd Floor, Crown House
Crown Street
Ipswich
Suffolk IP1 3HS
Tel: 0473 218951 Fax: 0473 231776

SURREY

FIN
3i plc
Guildford
Surrey
Tel: 0483 301773

SPARK
Surrey Research Park
PO Box 112
Guildford
Surrey GU2 5XL
Tel: 0483 579693 Fax: 0483 68946

TEC
Technology House
48–54 Goldsworth Road
Woking
Surrey GU21 1LE
Tel: 0483 728190 Fax: 0483 755259

TEC
AZTEC
Manorgate House, 2 Manorgate Road
Kingston-upon-Thames
Surrey KT2 7AL
Tel: 081 547 3934 Fax: 081 547 3884

UDIL
University of Surrey
Bureau of Industrial Liaison
Guildford
Surrey GU2 5XH
Tel: 0483 571281 Fax: 0483 300803

SUSSEX

FIN
3i plc
Brighton
Sussex
Tel: 0273 23164

TAYSIDE

DA
Scottish Enterprise, Tayside
Enterprise House
45 North Lindsay Street
Dundee
Tayside DD1 1NT
Tel: 0382 23100 Fax: 0382 201319

FIN
Tayside Enterprise Board
Fulton Road
Wester Gourdie
Dundee
Tayside DD2 4SW
Tel: 0382 621030 Fax: 0382 621014

UDIL
University of Dundee
Industrial Liaison Office
Tower Building
Dundee
Tayside DD1 4HN
Tel: 0382 25468 Fax: 0382 201604

TYNE & WEAR

FIN
3i plc
Newcastle
Tyne & Wear
Tel: 091 281 5221

GOV
DTI North East
Stanegate House
2 Groat Market
Newcastle
Tyne & Wear NE1 1YN
Tel: 091 232 4722

TEC
Wearside TEC
Derwent House
New Town Centre
Washington
Tyne & Wear NE38 7ST
Tel: 091 416 6161 Fax: 091 415 1093

TEC
Tyneside TEC
Moongate House, 5th Ave. Business Park
Team Valley Trading Estate
Gateshead
Tyne & Wear NE11 0HF
Tel: 091 487 5599 Fax: 091 482 6519

UDIL
University of Newcastle upon Tyne
Research & Industry Liaison
6 Kensington Terrace
Newcastle
Tyne & Wear NE1 7RU
Tel: 091 222 6091 Fax: 091 222 6229

WARWICKSHIRE

TEC
Coventry & Warwickshire TEC
Brandon Court
Progress Way
Coventry
Warwicks CV3 2TE
Tel: 0203 635666 Fax: 0203 450242

SPARK
University of Warwick Science Park
Barclays Venture Centre
Sir William Lyons Road
Coventry
Warwickshire CV4 7EZ
Tel: 0203 418535 Fax: 0203 410156

UDIL
University of Warwick
Technology Transfer Office
Senate House
Coventry
Warwickshire CV4 7AL
Tel: 0203 523523 Fax: 0203 461606

WEST GLAMORGAN

DA
Welsh Development Agency
Swansea Industrial Estate
Fforestfach
Swansea
West Glamorgan SA5 4DL
Tel: 0792 561666

SPARK
Swansea Innovation Centre
University College of Swansea
Singleton Park
Swansea
West Glamorgan SA2 7PP
Tel: 0792 295556 Fax: 0792 295613

TEC
West Wales TEC
Orchard House
Orchard Street
Swansea
West Glamorgan SA1 5DJ
Tel: 0792 460355 Fax: 0792 456341

UDIL
University College of Swansea
Swansea Innovation Centre
Singleton Park
Swansea
West Glamorgan SA2 8PP
Tel: 0792 295556 Fax: 0792 295613

WEST MIDLANDS

FIN
Sumit Equity Ventures
4th Floor, Edmund House
12 Newhall Street
Birmingham
West Midlands B3 3ER
Tel: 021 200 2244 Fax: 021 233 4628

FIN
3i plc
Birmingham
West Midlands
Tel: 021 200 3131

FIN
Birmingham Technology
Aston Science Park, Love Lane
Aston Triangle
Birmingham
West Midlands B7 4BJ
Tel: 021 359 0981 Fax: 021 359 0433

GOV
DTI West Midlands
Ladywood House
Stephenson Street
Birmingham
West Midlands B2 4DT
Tel: 021 632 4111

SPARK
Aston Science Park
Aston Triangle
Love Lane
Birmingham
West Midlands B7 4BJ
Tel: 021 359 0981 Fax: 021 359 0433

SPARK
Birmingham Research Park
Institute of Research and Development
Vincent Drive
Birmingham
West Midlands B15 5SQ
Tel: 021 471 4977 Fax: 021 472 5738

TEC
Walsall TEC
5th Floor, Townend House
Townend Square
Walsall
West Midlands WS1 1NS
Tel: 0922 32332 Fax: 0922 33011

TEC
Dudley TEC
Dudley Court S., Waterfront East
Level Street
Brierley Hill
West Midlands DY5 1XN
Tel: 0384 485000 Fax: 0384 483399

TEC
Wolverhampton TEC
2nd Floor
30 Market Street
Wolverhampton
West Midlands WV1 3AF
Tel: 0902 311111 Fax: 0902 23669

TEC
Sandwell TEC
1st Floor, Kingston House
438/450 High Street
West Bromwich
West Midlands B70 9LD
Tel: 021 525 4242 Fax: 021 525 4250

TEC
Birmingham TEC
16th Floor, Metropolitan House
1 Hagley Road
Birmingham
West Midlands B16 8TG
Tel: 021 456 1199 Fax: 021 454 7255

UDIL
University of Birmingham
Industrial Liaison Office
Chancellor's Court, Edgbaston
Birmingham
West Midlands B15 2TT
Tel: 021 414 3881 Fax: 021 414 3850

WEST SUSSEX

TEC
Sussex TEC
Gresham House
12–24 Station Road
Crawley
West Sussex RH10 1HT
Tel: 0293 562922 Fax: 0293 26308

WEST YORKSHIRE

FIN
Yorkshire Enterprise
Elizabeth House
9–17 Queen Street
Leeds
West Yorkshire LS1 2TW
Tel: 0532 420505 Fax: 0532 420266

FIN
3i plc
Leeds
West Yorkshire
Tel: 0532 430511

GOV
DTI Yorkshire & Humberside
Priestley House
3–5 Park Row
Leeds
West Yorkshire LS1 5LF
Tel: 0532 443171

SPARK
Listerhills Science Park
University of Bradford
Bradford
West Yorkshire BD7 1DP
Tel: 0274 733466 Fax: 0274 720910

TEC
Leeds TEC
Fairfax House
Merrion Street
Leeds
West Yorkshire LS2 8JU
Tel: 0532 446181 Fax: 0532 438126

TEC
Wakefield TEC
Grove Hall
60 College Grove Road
Wakefield
West Yorkshire WF1 3RN
Tel: 0924 299907 Fax: 0924 201837

TEC
Bradford & District TEC
Fountain Hall
Fountain Street
Bradford
West Yorkshire BD1 3RA
Tel: 0274 723711 Fax: 0274 370980

TEC
Calderdale & Kirklees TEC
Park View House
Woodvale Office Park
Brighouse
West Yorkshire HD6 4AB
Tel: 0484 400770 Fax: 0484 400672

TEC
Bradford & District TEC
5th Floor, Provincial House
Tyrell Street
Bradford
West Yorkshire BD1 1NW
Tel: 0274 723711 Fax: 0274 370980

UDIL
University of Leeds
Industrial Services
175 Woodhouse Lane
Leeds
West Yorkshire LS2 3AR
Tel: 0532 333444 Fax: 0532 445270

UDIL
University of Bradford
Industrial Liaison Services
Bradford
West Yorkshire BD7 1DP
Tel: 0274 733466 Fax: 0274 305340

WILTSHIRE

TEC
Wiltshire TEC
The Bora Building, Westlea Campus
Westlea Down
Swindon
Wiltshire SN5 7EZ
Tel: 0793 513644 Fax: 0793 542006

INDEX